# Why Are Things the Way They Are?

*Vignettes in Physics*
*A series by G. Venkataraman*

*Published*
The Many Phases of Matter
Chandrasekhar and His Limit

*Forthcoming*
Bose and His Statistics
At the Speed of Light
A Hot Story
The Quantum Revolution (3 vols)
   1. The Breakthrough
   2. QED: The Jewel of Physics
   3. What is Reality?

*Vignettes in Physics*

# Why Are Things the Way They Are?

## G. Venkataraman

Universities Press

© Universities Press (India) Private Limited 1992

First Published 1992
ISBN 0 86311 312 5

*Distributed by*
**Orient Longman Limited**

*Registered Office*
3-6-272 Himayatnagar, Hyderabad 500 029 (A.P.), India

*Other Offices*
Kamani Marg, Ballard Estate, Bombay 400 038
17 Chittaranjan Avenue, Calcutta 700 072
160 Anna Salai, Madras 600 002
1/24 Asaf Ali Road, New Delhi 110 002
80/1 Mahatma Gandhi Road, Bangalore 560 001
3-6-272 Himayatnagar, Hyderabad 500 029
Birla Mandir Road, Patna 800 004
Patiala House, 16-A Ashok Marg, Lucknow 226 001
S.C. Goswami Road, Panbazar, Guwahati 781 001

*Cover*: The mountain is intended to draw attention to the question: Why don't we have mountains very much higher than the Everest? See Chapter 5.

*Frontispiece*: Victor Frederick Weisskopf

*Typeset by* Alliance Phototypesetters, Pondicherry 605 013
*Printed in India by* offset at Pragati Art Printers, Hyderabad 500 004

*Published by* Universities Press (India) Private Limited
3-5-820 Hyderguda, Hyderabad 500 029

# Contents

| | |
|---|---|
| *Preface* | vii |
| What it is all about | 1 |
| On fundamental constants | 11 |
| How big is an atom? | 25 |
| On solids and liquids | 45 |
| Height of mountains | 62 |
| On the way to the stars | 68 |
| Something about stars | 81 |
| Cosmic numbers | 98 |
| Some parting thoughts | 109 |
| *Suggestions for further reading* | 112 |

# Preface

## To the adult reader

This book and others in this series written by me are inspired by the memory of my son Suresh who left this world soon after completing school. Suresh and I often used to discuss physics. It was then that I introduced him to the celebrated *Feynman Lectures*.

Hans Bethe has described Feynman as the most original scientist of this century. To that perhaps may be added the statement that Feynman was also the most scintillating teacher of physics in this century.

The Feynman Lectures are great but they are at the textbook level and meant for serious reading. Moreover, they are a bit expensive, at least for the average Indian student. It seemed to me that there was scope for small books on diverse topics in physics which would stimulate interest, making at least some of our young students take up later a serious study of physics and reach for the Feynman as well as the Landau classics.

Small books inevitably remind me of Gamow's famous volumes. They were wonderful, and stimulated me to no small extent. Times have changed, physics has grown and we clearly need other books, though written in the same spirit.

In attempting these volumes, I have chosen a style of my own. I have come across many books on popular science where elaborate sentences often tend to obscure the scientific essence. I have therefore opted for simple English, and I don't make any apologies for it. If a simple style was good enough for the great Enrico Fermi, it is also good enough for me. I have also employed at times a chatty style. This is deliberate. Feynman uses this with consummate skill, and I have decided to follow in his footsteps (whether I have succeeded or not, is for readers to say). This book is meant to be read for fun and excitement. It is a book you can even lie down in bed and read, without going to sleep I hope!

Naturally I have some basic objectives, the most important of which is to stimulate the curiosity of the reader. Here and there the reader may fail to grasp some details, and in fact I have deliberately pitched things a bit high on occasions. But if the reader is able to experience at least in some small measure the *excitement* of science, then my purpose would have been achieved. Apart from excitement, I have also tried to convey that although we might draw boundaries and try to compartmentalise

Nature into different subjects, she herself knows no such boundaries. So we can always start anywhere, take a random walk and catch a good glimpse of Nature's glory. Where she is concerned, all topics are 'fashionable'. There is today an unnecessary polarization of the young towards subjects that are supposed to be fashionable. To my mind this is unhealthy, and I have tried to counter it.

This series is essentially meant for the curious. With humility, I would like to regard it as some sort of a 'Junior Feynman Series', if one might call it that. With much love, and sadness, it is dedicated to the memory of Suresh who inspired it.

**To the young reader**

This book is slightly different from the others in this series. You will find many more calculations, but their inclusion is deliberate. Let me explain why I have done this.

I suppose you know that the size of an atom is roughly about $10^{-8}$ cm. Have you ever wondered why it is not something different, say, 1 cm? The height of Mount Everest is about 10 km. Why don't we have mountains 100 km high? In short, why are things the way they are? This is the question we deal with here.

To calculate accurately the size of an atom is not all that easy. However, if one wants merely to *estimate* the size, then that is a different matter altogether. In fact, if one is clever, then such estimates can be made in just a few steps. The calculations can even be scribbled on the back of an envelope! Physicists are very fond of such quickies, and they are quite useful too. This book gives you an introduction to them. It is fun to be able to do such rapid estimates.

**Acknowledgements**

As earlier, I am indebted to Professors V. Balakrishnan and N. Mukunda for a careful scrutiny of the manuscript and making many useful suggestions. Mr. A. Ratnakar and Mr. John D. Vincent rendered invaluable help in gathering reference material. Special thanks are due to CERN for providing a photograph of Professor Weisskopf and various pictures of CERN, besides granting permission to reproduce them. Mrs. Naga Nirmala rendered useful assistance with manuscript preparation. As always it has been a pleasure to work with Universities Press, and the cooperation of the editorial staff in particular is much appreciated.

G.VENKATARAMAN

# 1 What It Is All About

You must have heard of Geneva. It is a beautiful city in Switzerland, right on the border with France (Fig. 1.1). In that city is a famous laboratory called CERN which is the abbreviation for Organisation Européenne pour la Recherche Nucléaire, and is the French for the European

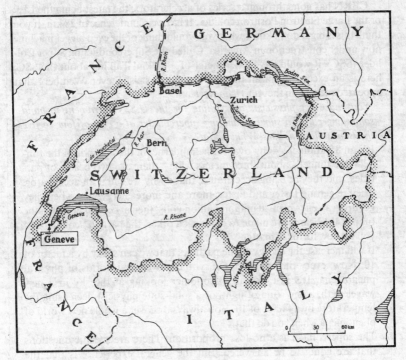

**Fig. 1.1** Map of Switzerland showing the location of Geneva.

Organisation for Nuclear Research. This lab has been set up jointly by many countries of Western Europe, and they have here many powerful particle accelerators for studying elementary particles (see Box 1.1). Here it was that the W boson predicted by Abdus Salam and Steven Weinberg was discovered. Carlo Rubbia who made the discovery won the

**Box 1.1** The first accelerators (these were built in the thirties) were quite small and could be accommodated in laboratory rooms. As the energy to which particles were accelerated increased, the accelerator size also started increasing. Today, the largest machines are several kilometres long and so they are buried below the ground since land for them is not available.

Typically, the machines are of the collider type. In such a machine, particles previously accelerated to the desired energy are injected into a storage ring where they circulate for hours, maybe even a few days. There is *ultra-high vacuum* in the ring, and the orbits are *very precise*. There are always two streams going in opposite directions, one of particles and the other of antiparticles. Their orbits are arranged to intersect at select spots where collisions can occur. These collisions produce all kinds of events which scientists then detect and study.

CERN has gone through a series of accelerators, the latest being the LEP or the Large Electron Positron collider. It has a circumference of 28 km. If you think this is big, I don't know what you would think of the even bigger machine now under construction in America. Called the Superconducting Super Collider (SSC), it would have a perimeter of ~ 85 km! Talking about the SSC, here is an excerpt from an interview with physicist Steven Weinberg that appeared recently in the *American Review*:

'You and other physicists are calling for the U.S. Government to build something called a superconducting supercollider at a cost of four to five thousand million dollars. What is that?

It's an 85-kilometer-long oval tunnel underneath the ground. In the tunnel—the tunnel's about three meters wide—there are two beams of particles, and they're being accelerated; that's why we call it an accelerator. They're getting faster and faster, more and more energy, and when they get up to an energy of 20 million million volts, then at certain points in the tunnel, the beams are made to cross each other, the particles collide, and you get new forms of matter.

It will increase the energy of the particles that we can study by a factor of 10. Now, every time you do that, you discover new worlds of physical phenomena. It's just like an astronomer looking at the sky in a new wavelength, a new kind of light, or a solid-state physicist decreasing the temperature by a factor of 10. You always discover a whole new world of phenomena when you do that.

The supercollider is a no-lose proposition. There are specific questions that are going to be answered. But the general experience is that the questions that you know are going to be answered never turn out to be the most important ones. The most important ones are things that you haven't thought of. It's opening up a new world. It's not a question of improving the efficiency of our work; it's a question of allowing the work to be done at all. Do you want to know the underlying laws of physics? Well, this is the way you have to go about it. I'm not saying that everything in physics is waiting for the supercollider. There are a lot of things that can be done in physics without the supercollider. But there's a certain kind of question—the question "Why? Why? Why?" getting down to the bottom of the chains of explanation—that can't be answered without this kind of facility. There's no argument at all in the community of American particle physicists that this is the right accelerator to build.'

Will accelerators keep getting bigger and bigger? At one time people thought there was no end to this business, and Enrico Fermi used to joke that one day the accelerator would become so big that it would have to encircle the globe. Fermi added that one-half of the world's population would be engaged in building, running and maintaining this machine, while the other half would be busy doing experiments with the machine! We are a long way off from that day but already many are worried, especially about the costs. Meanwhile, many clever ideas are being tried, and if there is a breakthrough, accelerators might shrink in size, at least for a while.

Nobel Prize. They are now building in CERN a great big accelerator with a circumference of 28 km. Electrons will be accelerated in this to nearly the speed of light, and they will go round and round, crossing between France and Switzerland millions of times every minute, without a passport, a visa, etc.!

Victor Weisskopf (see Box 1.2) was the Director of CERN in the sixties. He is a very learned and distinguished physicist who has made important contributions to atomic and nuclear physics. Every year in summer, high school students are invited to visit CERN, and one year, Weisskopf gave them some talks. This book is essentially based on those lectures. Weisskopf's title was *Modern Physics from an Elementary Point of View*, and he covered everything from atoms to the stars. The published lecture notes are rather brief and terse (in all, they run only to about 25 pages, and have six figures). But since the material covered by

**Box 1.2** Victor Frederick Weisskopf, popularly known as Vicky, was born in Austria in 1908. He received his doctorate from the University of Gottingen in 1934, and spent some time thereafter with Neils Bohr in Copenhagen and Pauli in Zurich. In 1937 he went to the U.S. where, during the war years he worked on the atomic bomb project. After the war he became Professor at M.I.T. From 1961 to 1965 he was the Director of CERN.

Weisskopf has made important contributions to quantum electrodynamics and to nuclear physics. Along with Wigner, he studied how spectral lines get broadened. Perhaps Weisskopf would be best remembered as one who explains things simply and exceedingly well. His books are very popular. His *Theoretical Nuclear Physics* written in 1952 is still an excellent book to learn from. For the layman, he has written one called *Knowledge and Wonder*.

Weisskopf is very interesting, I have decided to expand his brief notes; and this book is the result. In essence, it is Weisskopf's lectures retold.

To understand this book, you really don't need anything more than high school physics. Of course here and there you might need something more but, as hinted previously, I have introduced whatever extra is required.

I want you to notice two things. One of course is how simple and brief every calculation is; this I have already mentioned (in the introductory note entitled 'To the young reader'). The other and the more important thing is that in almost all cases (the sizes of things, for example) the answers are obtained mainly in terms of various fundamental constants.

There is yet another thing you should notice about these estimates which is that they are not supposed to be very accurate. They are intended only to give an *order of magnitude* value—that is to say, if we get an answer like 2.3 and the correct answer is 7.6, that is quite OK. The point is, the estimate does not come out as 0.0023 or 2300; it is just a few times bigger or smaller than the correct value (see Box 1.3).

---

**Box 1.3** According to the well-known Soviet physicist Ginzburg, 'One of the laws of cosmic physics is expressed by the equation $1 = 10$.' What Ginzburg means is that agreement to within an order of magnitude or so is excellent in any cosmological argument.

---

Making a back-of-the-envelope calculation is not all that simple. It is an art; but it can and *should* be cultivated. This book gives you examples of how such calculations are made. Once you master the art, you can have a lot of fun.

## The Accelerator Story

As mentioned in Box 1.1, over the years accelerators have been getting more and more powerful. The smaller the scale of length one wants to probe, the greater the energy and therefore the larger the machine. The picture and figures here give some idea of how big today's machines are, and what one does with them.

(a) Aerial view of the two-mile long Stanford linear accelerator.
(*Courtesy CERN PHOTO*)

# 6 Why are things the way they are?

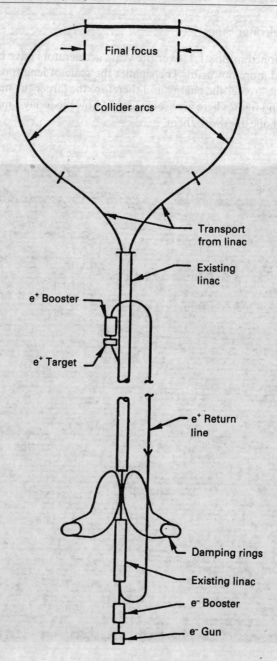

(b) A schematic drawing of the accelerator shown in (a). The drawing is inverted with respect to the photograph. (*Courtesy CERN PHOTO*)

(c) Aerial view of the accelerator at Fermilab (USA). (*Courtesy CERN PHOTO*)

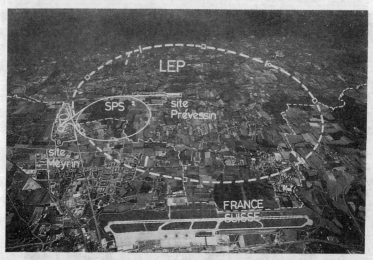

(d) Aerial view of CERN. In the foreground can be seen the runway of Geneva airport. The big circle shows the line of the LEP tunnel (~ 28 km circumference). Notice the boundary between France and Switzerland. (*Courtesy CERN PHOTO*)

(e) LEP tunnel during excavation. Depth is 45 metres below ground. (*Courtesy CERN PHOTO*)

(f) LEP collider tunnel. (*Courtesy CERN PHOTO*)

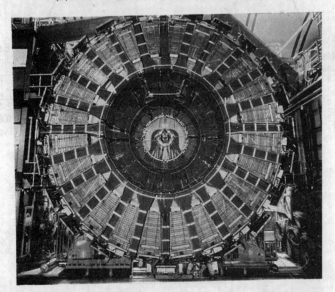

(g) An experimental station under construction. (*Courtesy CERN PHOTO*)

(h) Typical picture from a detector. (*Courtesy CERN PHOTO*)

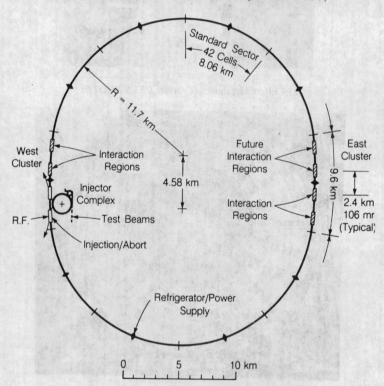

(i) Schematic of the Superconducting Super collider. (*Courtesy CERN PHOTO*)

# 2 On Fundamental Constants

This chapter contains material you may not be familiar with. Weisskopf does not discuss it; perhaps he takes it all for granted but I do not. A feature of this chapter is that the material in the boxes outweighs that in the main body of the chapter itself!

What is a fundamental constant? I don't know how to define it rigorously, and so I shall simply declare it to be a constant that occurs in physics and is important. Maybe this is not a very satisfactory definition, but very soon you will get an idea of what I mean. Perhaps an example would help. Let us take the famous law of gravitation. It says that the force of attraction between two objects with masses $m_1$ and $m_2$ is proportional to $(m_1 m_2)/r^2$. We know that the constant of proportionality is $G$, the gravitational constant. Well, $G$ is a fundamental constant. Another example is the charge on the electron; get the idea?

There are about a dozen or so important constants drawn from various domains of physics (as I shall explain presently). Some like the electronic charge $e$ and $\hbar$ (which is related to the Planck's constant $h$) represent basic units, that is to say charge and angular momentum come in multiples of $e$ and $\hbar$ respectively. $G$ on the other hand is a constant of proportionality. The velocity of light $c$ is a ceiling. The electron mass $m_e$ does not belong to any of the above categories; but because it denotes the mass of a fundamental particle like the electron, it also has a special status, i.e. it belongs to the club of fundamental constants. So the club has different types of members.

Earlier I said fundamental constants arise from various domains of physics. Let me explain that now. Take for example $G$. This comes from gravitation. Similarly, relativity contributes $c$ (for a quick primer on relativity, see Box 2.1). Thermodynamics and some of 19th century physics gave us the Avogadro number $N_A$, the Boltzmann constant $k_B$ and the gas constant $R$. Quantum mechanics brought in the Planck's constant $h$ while the subatomic world has contributed the electronic charge $e$, the electronic mass $m_e$, and the proton mass $m_p$. Finally, from the study of the Universe or the cosmos, we have received $M_\odot$, the mass of the Sun and $L$, the Hubble length. ($M_\odot$ is not really a fundamental constant in the same sense as, for example, $c$ or $e$ or $h$. It is more a

**Box 2.1** The theory of relativity is one of the finest achievements of the human mind. It was unfolded in two stages—first came the Special theory, followed then by the General theory. Both are due to the great Albert Einstein. For more details see the companion volume *At the Speed of Light* but here let me give you a capsule summary.

You must have learnt that the mass of an object is a constant and that only its weight varies as the object is taken to the Moon, Mars, etc. Relativity produced a surprise for it said that the *mass of an object varies with velocity*, the variation being

$$m = m_0 / \sqrt{(1 - v^2/c^2)}$$

See also figure (a). We notice that the mass variation is really not serious until the velocity $v$ approaches that of light. Incidentally, $v$ cannot be greater than $c$ for then the mass would become imaginary, which is not allowed. So the velocity of light is a ceiling—nothing can travel at a speed exceeding that. The quantity $m_0$ is called the *rest mass*, and is the mass the object has at zero velocity.

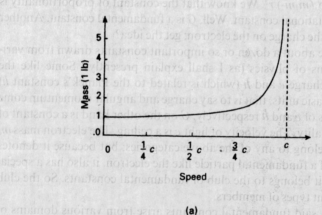

(a)

The second important effect predicted by the Special theory is the *slowing down of clocks*. Suppose you had a stop watch and ran with it at nearly the speed of light. Your watch would then slow down. What it means is that when your watch ticks one second on the dial, a man who is standing still and is watching you run will find that a stop watch he is carrying would have ticked off many more seconds—see figure (b).

Finally, there is the famous mass–energy relation $E = mc^2$. There was once a cartoon showing Einstein scribbling formulae on a blackboard. He first writes $E = ma^2$ and then cancels it. Next he tries $E = mb^2$ and scores that out

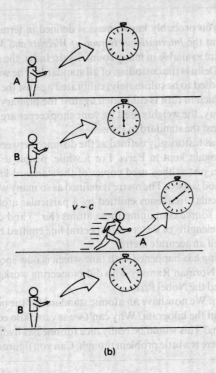

(b)

> too. He then comes up with $E = mc^2$ and shouts Eureka! What this famous formula says is that mass (even rest mass) can be converted into energy and that energy can be converted back into mass. Without this formula, there wouldn't have been any atomic bomb. The most common example of the reverse process namely, energy becoming mass is the so-called *pair production*. This refers to the disappearance of a photon (light quantum) and the appearance of an electron–positron pair. For this to happen, the photon must have an energy of at least 1 MeV which is the energy equivalent of the rest mass of an electron and a positron.

convenient constant.) Shortly, I shall give a list of the constants together with some comments. But first I must mention something else.

Many of the constants have dimensions, for example, the mass of the electron. You must be aware that there are standards for mass, length and time (see Box 2.2). So, where required, fundamental constants will be expressed in terms of these units. These days, everything is expressed in SI units (see Box 2.3). I am a bit old-fashioned, and so allow me to use units I am more familiar with namely, the CGS or the MKS units; I suppose you know what these refer to.

## 14   Why are things the way they are?

> **Box 2.2**   As you probably know, mass is defined in terms of the standard kilogram kept at the *International Bureau of Weights and Measures* in Paris. Copies of this are available in many countries. In India, the *National Physical Laboratory* in Delhi is the custodian of all standards. The weights used in our shops are supposed to be (ultimately) calibrated against the *secondary* standard in Delhi which in turn is calibrated against the primary standard kept in Paris. In this way, the weights used by your shopkeeper are (supposed to be) calibrated against the standard in Paris.
>
> The metre was historically defined as the distance between two marks on a platinum bar, again kept in Paris. For a while, people used copies of this standard metre, just as they used copies of the standard kilogram. Now an ingenious method is used. The metre is defined as so many wavelengths of the light of a particular frequency emitted by a particular atom. So if you can arrange a light source containing these atoms ($Kr^{86}$) and count the correct number of wavelengths of the correct spectral line emitted by this atom, you have got yourself an accurate metre!
>
> The same thing has happened with time which is now specified in terms of atomic clocks. Norman Ramsay who did pioneering work in this area, was recently awarded the Nobel Prize.
>
> One question: We now have an atomic standard for the metre and the second. What about the kilogram? Why can't we say one kilo equals the mass of so many protons? This would be really nice for one has hydrogen practically everywhere. There is a little problem though. Can you figure out what it is?

> **Box 2.3**   The term SI units comes from the French label Système International d' Unités. It has seven base units as given below:
>
> | Physical quantity | Name of unit | Symbol |
> |---|---|---|
> | Length | Metre | m |
> | Mass | Kilogram | kg |
> | Time | Second | s (or sec) |
> | Electric current | Ampere | A (or amp) |
> | Temperature | Kelvin | K (or °K) |
> | Light intensity | Candela | cd |
> | Amount of substance | Mole | mole |
>
> All physical quantities involve combinations of the above.

In passing I should mention that the derived quantities describing electromagnetic phenomena are in a class by themselves. Various systems of units are available for describing these, proposed at various times and with various intentions. Naturally, each system has its own advantage. When one encounters these different systems for the first time, they look rather confusing. Certainly I was confused on first encounter. (I am not sure if all the confusion has disappeared!). In Box 2.4, I introduce you briefly to the various systems; mind you, it is only an introduction and not an exposition. Just a 'Hey you, watch out!' kind of thing. One good thing. We won't be bothered with all this business here.

Tables 2.1 and 2.2 give some important fundamental constants. Of these the latter table contains a readily usable set (in CGS units). You will observe that some of them are dimensionless. In Chapter 8, such constants would be in the limelight.

---

**Box 2.4** Various systems of units are used for describing electromagnetic quantities, examples being the electrostatic units, electromagnetic units, Gaussian units, Heaviside units and rationalised MKSA units. The differences between these are not quite like those between the CGS and the (now practically extinct) Foot-pound-second system. In the latter, somebody picked a hunk of matter and said: 'I don't accept the kilogram. This is my standard mass'. The same happened for the unit of length. The differences we are now talking about are quite different. In a sense, the differences arise due to the various choices made for certain constants which appear in some of the formulae for electromagnetic quantities. Two such are the electric permittivity $\varepsilon_0$ and the magnetic permeability $\mu_0$. Historically, one system of units was derived by taking $\varepsilon_0 = 1$; this was the electrostatic system (ESU). Similarly, by setting $\mu_0 = 1$, one obtained the electromagnetic system of units (EMU). This was certainly convenient for dealing with electrostatics and magnetostatics respectively. But there were problems, and so other systems were invented.

A comprehensive way of describing these various systems would be to draw up a table as below.

| System    | $k_1$         | $k_2$          | $\alpha$ | $k_3$ |
|-----------|---------------|----------------|----------|-------|
| ESU       | 1             | $1/c^2$        | 1        | 1     |
| EMU       | $c^2$         | 1              | 1        | 1     |
| Gaussian  | 1             | $1/c^2$        | $c$      | $1/c$ |
| Heaviside | $1/4\pi$      | $1/(4\pi c^2)$ | $c$      | $1/c$ |
| RMKSA     | $10^{-7} c^2$ | $10^{-7}$      | 1        | 1     |

The constants $k_1$, $k_2$, $\alpha$ and $k_3$ above occur in some of the equations of electromagnetism (like Faraday's laws, for example). RMKSA stands for the rationalised MKS system, with the Ampere as the unit of current. One must be careful in going over from one system to another. For example, the unit of capacitance in the RMKSA system is the Farad; in the Gaussian system it is cm.

# 16  Why are things the way they are?

**Table 2.1** A comprehensive list of fundamental constants

| Quantity | Symbol | Num. Value | Units(SI)* |
|---|---|---|---|
| Speed of light in vacuum | $c$ | 299792458 | m·s$^{-1}$ |
| Fine structure const. | $\alpha$ | 7.2973506 | |
| Reciprocal fine structure const. | $1/\alpha$ | 137.03604 | |
| Elementary charge | $e$ | 1.6021892 | $10^{-19}$ C |
| Planck const. | $h$ | 6.626726 | $10^{-34}$ J.s |
| $(h/2\pi)$ | $\hbar$ | 1.054589 | $10^{-34}$ J.s |
| Avagadro const. | $N_A$ | 6.022045 | $10^{-23}$ mol$^{-1}$ |
| Atomic mass unit | $u$ | 1.6605655 | $10^{-27}$ kg |
| Electron rest mass | $m_e$ | 9.109534 | $10^{-31}$ kg |
| Proton rest mass | $m_p$ | 1.6726485 | $10^{-27}$ kg |
| Neutron rest mass | $m_n$ | 1.6749543 | $10^{-27}$ kg |
| Elementary charge to mass ratio | $(e/m_e)$ | 1.7588047 | $10^{11}$ C·kg$^{-1}$ |
| Rydberg const. | $R_\alpha$ | 1.097373177 | $10^7$ m$^{-1}$ |
| Bohr radius | $a_0$ | 5.2917706 | $10^{-11}$ m |
| Bohr magneton ($e\hbar/2m_e$) | $\mu_B$ | 9.274078 | $10^{-24}$ J·T$^{-1}$ |
| Electron magnetic moment | $\mu_e$ | 9.284832 | $10^{-24}$ J·T$^{-1}$ |
| Proton magnetic moment | $\mu_p$ | 1.4106171 | $10^{-26}$ J·T$^{-1}$ |
| Nuclear magneton ($e\hbar/2m_p$) | $\mu_N$ | 5.050824 | $10^{-27}$ J·T$^{-1}$ |
| Compton wavelength electron ($h/m_e c$) | $\lambda_c$ | 2.4263089 | $10^{-12}$ m |
| proton ($h/m_p c$) | $\lambda_{c,p}$ | 1.3214099 | $10^{-15}$ m |
| Boltzmann const. | $k_B$ | 1.380662 | $10^{-23}$ J·K$^{-1}$ |
| Gravitational const. | $G$ | 6.6720 | $10^{-11}$ m$^3$·s$^{-2}$·kg$^{-1}$ |

*
C – Coulomb
J – Joule
T – Tesla
Other symbols are defined in Box 2.3

## Table 2.2

| Constant | Symbol | Value |
|---|---|---|
| Planck's const. | $h$ | $6.62 \times 10^{-27}$ erg sec |
|  | $\hbar$ | $1.054 \times 10^{-27}$ erg sec |
| Electron volt | eV | $1.602 \times 10^{-12}$ ergs |
| Rydberg | Ry | 13.6 eV |
| Fine structure const. | $\alpha$ | 1/137.04 |
| Electron mass | $m_e$ | $0.91 \times 10^{-27}$ gm |
| Proton mass | $m_p$ | $1.67 \times 10^{-24}$ gm |
|  |  | $\sim 1836\, m_e$ |
| Classical electron radius | $r_0$ | $2.82 \times 10^{-13}$ cm |
| Unit of charge | $e$ | $4.8 \times 10^{-10}$ esu |
| Compton wave length | $\lambda_c$ | $3.86 \times 10^{-11}$ cm |
| Avagadro no. | $N_A$ | $6.02 \times 10^{23}$ /mole |
| Boltzmann const. | $k_B$ | $1.38 \times 10^{-16}$ erg/°K |
| Gas const. | $R$ | $8.31 \times 10^7$ |
| Gravitational const. | $G$ | $6.67 \times 10^{-27}$ cm$^3$/gm.sec$^2$ |
| Solar mass | $M_\odot$ | $1.99 \times 10^{33}$ gm |
| Hubble length | $L$ | $\sim 10^{27}$ cm |

Now for some comments.
1. $e$ denotes the fundamental unit of charge. The charge on an electron is taken to be $-e$, and that on a proton as $e$ (which is the same as $+e$).
2. $h$ is the Planck's constant. Its existence became known when Planck proposed his famous quantum theory around 1900. Planck's tombstone bears the simple epitaph:
$$h = 6.62 \times 10^{-27} \text{ erg sec (see Box 2.5).}$$

---

**Box 2.5.** The quantum revolution may be said to have started with the discovery by Max Planck of the Planck radiation formula. One of the important problems studied in the second half of the 19th century was the radiation spectrum of a blackbody. Several efforts to explain the observations were made but they all failed until finally Planck succeeded in 1900. Planck's formula says that the energy density $\rho(\nu, T)$ is given by

$$\rho(\nu, T) = \frac{8\pi h \nu^3}{c^3} \frac{1}{\exp(h\nu/k_B T) - 1}$$

where $k_B$ is the Boltzmann constant, $h$ is the Planck's constant, $\nu$ is the frequency, and $T$ is the absolute temperature. Planck discovered that the energy levels of an oscillator are quantised. Later, Einstein discovered that radiation energy also is quantised. Neils Bohr applied the quantisation idea to atomic energy levels, thus paving the way for quantum mechanics.

3. $\hbar = h/2\pi$ is a unit of angular momentum (see Box 2.6). Angular momentum in Nature occurs in multiples of $\hbar$, i.e. one has $\hbar$, $2\hbar$, $3\hbar$, ..., etc. $\hbar$ is read as $h$ bar or $h$ cross.

> **Box 2.6** You probably know what is meant by momentum and also that force is equal to the rate of change of momentum. Forget momentum for a second. Let us say you want to turn a screw. You would of course use a screwdriver. Sometimes it is easy to turn but sometimes it is difficult. And when it is so, you apply extra force. That is really an *angular force* called *torque*. This torque is in turn the rate of change of something called the *angular momentum*. So,
>
> linear momentum ⟷ linear force,
>
> angular momentum ⟷ angular force.
>
> By the way, both the momenta are vectors but there is an interesting difference between them. You should ask your teacher or a senior to explain this difference.
>
> The magnitude of the linear momentum is given by $mv$. In a similar fashion, the magnitude of angular momentum is given by $mvr$ where $r$ is the radius of the orbit in which the particle is moving.
>
> Consider now a big object like a flywheel. We would think that by giving it various angular velocities, we can make the flywheel have any angular momentum we please. Not true! Angular momentum can only be increased in discrete steps of $\hbar = h/2\pi$. Of course, $\hbar$ is so very tiny that in the case of the flywheel the step increase is hardly visible and we think (mistakenly of course) that the angular momentum can vary continuously. But step into the world of atoms, and it is a different story, as Niels Bohr discovered for the first time. More about that in Chapter 3.

4. $m_e$ is the mass of the electron, sometimes called the electron *rest mass*. Associated with it, on the basis of the special theory of relativity, is the *rest energy* of the electron given by $m_e c^2$.

5. $N_A$ is the Avogadro number, and denotes the number of atoms in one gram molecule of the substance. You must have come across this constant in chemistry.

6. The Boltzmann constant was introduced by Max Planck who named it after Boltzmann. It occurs in the famous Boltzmann formula for entropy (see Box 2.7). We will often encounter the combination $k_B T$, where $T$ is the absolute temperature, i.e. temperature in degrees Kelvin. The combination $k_B T$ obviously has the dimensions of energy.

**Box 2.7** The Boltzmann constant $k_B$ is related to a quantity called the *entropy*. The concept of entropy was amplified by Ludwig Boltzmann during the last century, and may be described as one of the great discoveries of physics. The simplest way of explaining entropy is for me to quote the great mathematical physicist Freeman Dyson. He asks: 'What is heat?' and then proceeds to give the following answer:

'Heat is disordered energy. So with two words, the nature of heat is explained... Energy can exist without disorder. For example, a flying rifle bullet or an atom of $U^{235}$ carries ordered energy. The motion of the bullet is of a kind we call kinetic. When the bullet hits a steel plate and is stopped, the energy of its motion is transferred to the random motions of the atoms in the bullet and in the plate. The disordered energy makes itself felt in the form of heat... We measure heat precisely in terms of numbers... it is clear that to specify heat we must use at least two numbers; one to measure the quantity of heat, and the other to measure the quantity of disorder. The quantity of heat is measured in terms of a practical unit called the calorie... The quantity of disorder is measured in terms of a mathematical quantity called entropy.'

Boltzmann gave a method for calculating the entropy. In modern language it can be reduced to the formula

$$S = k_B \log_e W$$

Here $S$ is the entropy, and $W$ a quantity related to the random motions. Boltzmann of course indicated how to get at $W$. Once that is known, $S$, the entropy comes out of the above formula. $k_B$ is the constant of proportionality, and is referred to as the Boltzmann constant.

Boltzmann's life ended in a tragedy. Practically single handed, he developed the famous kinetic theory of gases. As you probably know, it is based on the assumption that gases are made up of flying molecules and atoms. Some people did not believe in atoms during those days, and they attacked Boltzmann viciously. Unable to bear the hurt, Boltzmann ended his life by jumping into the Adriatic sea. Not far from that place is located the famous International Centre for Theoretical Physics founded by Abdus Salam. The Boltzmann formula now adorns his tombstone.

7. eV is an electron volt, and its meaning is explained in Fig. 2.1. It is a unit of energy. In atomic physics and in solid state physics, eV is a convenient unit of energy to use. In nuclear physics, the energies one deals with are in the range of millions of electron volts. So it is common to use the unit MeV which denotes $10^6$ eV. In high-energy physics, people find that MeV is too small and prefer GeV (=1000 MeV) or even TeV (= 1,000,000 MeV). See Box 2.8.
8. Rydberg is another unit of energy, often used in atomic physics.

## 20  Why are things the way they are?

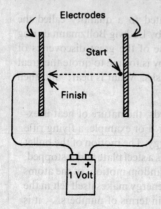

**Fig. 2.1** Suppose an electron is moved up against an electric potential as shown in the figure. Work has to be done for this purpose, just as one has to do work to lift a stone against gravity. The work done in moving an electron against a potential of 1 volt is referred to as an *electron volt*.

---

**Box 2.8**   In physics one often comes across a wide range of the same physical quantity. This is not surprising since physics deals with objects as tiny as an electron, and as large as galaxies. Powers of ten come in handy in such situations. When dealing with fractions one uses negative powers, while positive powers are used for dealing with large quantities. The special prefixes and suffixes used in this connection are summarised below:

Prefixes used for multiples and submultiples of metric quantities.

| | | | |
|---|---|---|---|
| $10^{-1}$ | deci | $10^{1}$ | deca |
| $10^{-2}$ | centi | $10^{2}$ | hecto |
| $10^{-3}$ | milli | $10^{3}$ | kilo |
| $10^{-6}$ | micro | $10^{6}$ | mega |
| $10^{-9}$ | nano | $10^{9}$ | giga |
| $10^{-12}$ | pico | $10^{12}$ | tera |

---

Rydberg is the name of a Swedish spectroscopist. 1 Rydberg = $(m_e e^4/2\hbar^2)$.

9. The Ångstrom (Å) is a unit of length, one Ångstrom being equal to $10^{-8}$ cm. This unit too is named after a Swedish spectroscopist, and is often used in atomic physics, optics and spectroscopy.

10. Fermi also is a unit of length and is equal to $10^{-13}$ cm. It is often used in nuclear physics, and is named after the famous Italian physicist Enrico Fermi who, among other things, built the first nuclear reactor and 'tamed' nuclear energy.

11. The Hubble length (see Box 2.9 ) is unlike the other constants in that its value is a guess!

---

**Box 2.9** Hubble was a great astronomer. It was he who first provided evidence that our Universe is expanding. Hubble established this by looking at the light coming from distant galaxies. If the Universe is expanding, then the galaxies would all appear to be running away from us. There is a simple way of understanding this. Imagine an ant on a rubber balloon which is being inflated. The ant is stationary and so are a few pieces of chalk dust sprinkled around the ant. However, because of the expansion of the balloon, the ant would think that the chalk dust specks around it are all running away. In the case of our Universe, it was found that

Recession velocity = constant × distance of the galaxy

The constant in the above relation is denoted by $H$, and is referred to as the *Hubble constant*. Its numerical value is estimated to be 20 km/sec/$10^6$ light years. The Hubble length is given by $(c/H)$. It is a cosmic yardstick. It is the distance at which the recession velocity of a galaxy equals the velocity of light. The value of $H$ is a bit of a guess; the one quoted is the best estimate available at present.

---

12. The fine structure constant $\alpha = (e^2/\hbar c)$ is quite interesting. It is dimensionless, and combines the atomic world with the world of electromagnetic radiation. So, whenever we deal with radiation from atoms, we will encounter this constant (see Box 2.10).
13. The classical electron radius is so tiny a quantity that one may

---

**Box 2.10** I heard this story from a good friend of mine who is an excellent physicist and also a superb teacher. When he was a student in America about twenty years ago, he and his friend once attended a summer school on quantum electrodynamics in a place called Rhode Island. They drove to Rhode Island from Boston in their own car. (Those days, second-hand cars were often quite cheap and even students could afford them. I don't know about now.) On the highway, there was a tiny Volkswagen car tailing them. When they left the highway to go to Rhode Island, the Volkswagen did likewise. This friend of mine became quite curious and looked back. By now the Volks was quite close, and my friend could see that the license plate number was the same as the reciprocal of the accepted value of the fine structure constant. Both he and his friend were quite surprised but thought that the choice of this number must have been an accident and that the driver of the Volks was quite ignorant about the significance of this number.

When these students finally reached their destination and parked their car, they found that the Volks was also swinging into the same parking lot. Unable to contain their curiosity, they went near that car to see who the driver was. Guess who stepped out? None other than Julian Schwinger, one of the Trinity of quantum electrodynamics! In 1965, Schwinger shared the Nobel Prize with Feynman and Tomonaga for their contributions to the theory of quantum electrodynamics.

genuinely wonder how such a small quantity has a place in classical physics. The story goes like this. J.J.Thomson discovered the electron in 1897. At that time, nobody dreamt about quantum effects. Thomson simply asked: Suppose I try to hit one electron with another. How big does the electron appear? He calculated this quantity and found that the electron behaved as if it had a radius ($e^2/m_e c^2$). Notice that this expression does not involve $h$, which was not known at that time anyway. Only the quantities known to classical physics are involved.

14. The French scientist de Broglie discovered that particles also have a wavelike character (see Box 2.11). The Compton wavelength may be regarded as the de Broglie wavelength of an electron travelling close to the speed of light.

Table 2.3 gives a few useful conversion factors.

**Table 2.3**

| To convert from | To | Multiply by |
| --- | --- | --- |
| Nanometres | Ångstroms | 10 |
| Ångstroms | Nanometres | 0.1 |
| Pascal | Torr | 0.0075 |
| Torr | Pascal | 133.3 |
| Atmosphere (bar) | Torr | 760 |
| Torr | Atmosphere | 0.001316 |
| Newton | Dyne | $10^5$ |
| Dyne | Newton | $10^{-5}$ |
| eV | Erg | $1.602 \times 10^{-12}$ |
| Erg | eV | $6.24 \times 10^{11}$ |
| eV | Kilo joules/mole | 96.49 |
| Joules | Calories | 0.239 |
| Calories | Joules | 4.184 |

One last word before I wind up this chapter. It is good to memorise the values of the fundamental constants up to one or two decimal places. Actually, it is not all that difficult. Physicists who are active use these constants so often that they know these numbers by heart. They may

**Box 2.11** Louis de Broglie was a French physicist. In 1924, he suggested that every particle has associated with it a wavelike nature. If $v$ is the velocity of the particle and $m$ its mass, then the wavelength $\lambda$ of the associated de Broglie wave (or matter wave as it is sometimes called) is given by

$$\lambda = (h/mv)$$

Davisson and Germer in America and G.P.Thomson in England decided to verify if this was true. They experimented with electrons and said that if electrons behaved like waves, then there must be diffraction effects; and sure enough there were. Naturally they won the Nobel Prize. So of course did de Broglie. By the way, G.P. Thomson was the son of J.J. Thomson who discovered the electron. The father got the prize for discovering the particle, and son for proving that it was also a wave!

Let us imagine that we superpose many wavetrains as in figure (a) in a suitable manner (i.e. with properly adjusted phases and amplitudes). Then, except in a small region, these wavetrains cancel each other. What is left is called a *wave packet*. The size of a wave packet is roughly given by de Broglie's formula. It means that the particle would appear like a fuzzy ball of diameter ~ $\lambda$. Notice that as $v$ increases, the fuzziness decreases. (The symbol ~ means *of the order of*.)

not remember telephone numbers, names of people, etc., but they certainly would remember their constants. In fact, they often use these numbers for their combination locks. So if you want to secretly open the combination lock of a good physicist, try the fundamental constants.

## 24  Why are things the way they are?

Richard Feynman actually did this, and narrates the story in his racy autobiography: *Surely you're joking, Mr. Feynman*. If you have not already read it, you have really missed something. Get the book and read it as soon as you can. Having read it, remember that behind all that humour, Feynman was a damn serious physicist. Serious physicists are always making quick estimates while thinking. And making quick estimates is what this book is all about.

# 3 How Big Is An Atom?

## 3.1 The hydrogen atom

We start with the world of atoms. Among the atoms, hydrogen is the simplest; it has one proton, around which revolves an electron. You must have heard of the Bohr atom model (see Box 3.1), the model given by Niels Bohr (see Box 3.2) before the advent of quantum mechanics. According to it, electrons are supposed to go round and round the nucleus like planets do around the Sun. These orbits are sometimes referred to as the Bohr orbits.

---

**Box 3.1** In the last century, people believed that atoms were the ultimate building blocks of matter. They did not imagine that atoms could have an internal structure of their own. In 1897, J.J.Thomson discovered the electron and proved that the atom was in fact divisible. Since the atom as a whole was electrically neutral, it was clear that the atom must also contain a positive charge. For a while, people did not know where inside the atom this positve charge was sitting, until Rutherford gave the answer in 1911. By his brilliant experiments he showed that the positive charge was concentrated in a tiny spot called the nucleus. Next came the question: How are the electrons organised with respect to the nucleus? Niels Bohr answered that one in 1913.

Let $Ze$ be the charge on the nucleus, and suppose that the electron is going round the nucleus in an orbit of radius $r$. Equating the attractive electrostatic force to the centrifugal force, we have

$$(Ze^2/r) = (m_e v^2/r)$$

The velocity $v$ and the radius $r$ are determined by the quantisation condition

$$mvr = n\hbar$$

What the above rule says is that the angular momentum of the electron must be an integral multiple of the basic quantum $\hbar$. Only those orbits for which this is true are permitted. This was Bohr's great discovery. Once he discovered this, the rest was simple. He found that the energy of the electron

---

in the nth orbit (integer $n$ is defined by the angular momentum quantisation) is given by

$$E_n = -[(Ze)^2/2n^2 a_0]$$

where $a_0 = (\hbar^2/m_e e^2)$ — see figure (a).

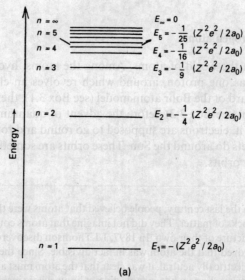

(a)

I am sure you know all this and must be wondering what it is that Weisskopf is trying to do, and how his treatment is any simpler. That would become clear when Weisskopf goes beyond the hydrogen atom. Those problems are a lot harder, and the Bohr model has no simple answers for them.

In hydrogen, the electron would normally be in the innermost Bohr orbit. When we ask the question: 'What is the size of the hydrogen atom?', we are really asking for the size of the first Bohr orbit.

For a moment, let us go back to planets and the Sun. The Sun attracts the planets, and the Sun being a very big object, planets should really fall into the Sun. But that has not happened; if it had, you and I wouldn't be here! Why hasn't the Earth fallen into the Sun? One way of understanding this is to say that it hasn't because there is an outward force acting on the Earth; this is the centrifugal force which arises because the Earth is actually whirling around the Sun. So the Earth is in a stable orbit under the balance between two forces: (i) gravity which pulls the

> **Box 3.2** Neils Bohr was born in Copenhagen, Denmark in 1885. His father was a Professor of Physiology and his brother a mathematician (also a very good football player). While still a student, he won a science contest organised by the Danish Academy of Sciences. The results were published as an original article in the *Transactions of the Royal Society*.
>
> In 1911 Bohr went to England. He first spent a year in Cambridge with Sir J.J.Thomson and then went to Rutherford's laboratory in Manchester. It was here that he developed the famous atom model, for which he received the Nobel Prize in 1922. Later, Bohr did some important work on nuclear fission. He would be remembered best for his interpretation of quantum mechanics, and for his efforts to convince Einstein about quantum uncertainty.

Earth towards the Sun, and (ii) centrifugal force which pushes the Earth away from the Sun. You may also say that there is a balance between the potential and the kinetic energies; that would be just another way of describing things. This balance determines the radius of the Earth's orbit; and it is different for the different planets, which is why they all have different orbits.

What is the sort of balance we must consider in the case of the atom? As in the solar system, it is a balance between the potential energy and the kinetic energy. Let us take these things one by one.

Let us say the electron is in an orbit of radius $r$. What is the potential energy of interaction between the electron and the proton? The two particles experience an attraction on account of electrostatic interaction as well as the gravitational interaction, and the two contributions are (see Box 3.3):

$$\text{Gravitational} \quad E_g = -Gm_p m_e/r \tag{3.1a}$$

> **Box 3.3** Many commonly occurring forces can be derived from suitable 'potentials'. Some illustrations.
>
> Gravitational force:
>
> ```
> 0─────────────0
> m₁            m₂
> ( ←────r────→ )
> ```
>
> Potential $V(r) = -G \cdot (m_1 m_2/r)$
>
> Force $F = -(dV/dr) = G \cdot (m_1 m_2/r^2)$
>
> This force is always attractive, i.e. the two masses always attract each other.

> Electrostatic force:
>
> $$0 \underset{q_1}{\quad\quad} \underset{q_2}{\quad\quad} 0$$
>
> Potential $V(r) = (q_1 q_2 / r)$
>
> Force $F = -(dV/dr) = -(q_1 q_2 / r^2)$
>
> If both the charges have the same sign, the force is repulsive, while if they have opposite signs it is attractive. Notice that whether the force is going to be attractive or repulsive is determined by the signs of the two charges.

$$\text{Electrostatic} \quad E_e = -e^2/r \quad\quad (3.1b)$$

If you put in numbers (I suggest you do this), you will find that the gravitational attraction is very much weaker than the electrostatic. So we can safely neglect the former.

BETWEEN TWO ELECTRONS

$$\frac{\text{Gravitation Attraction}}{\text{Electrical Repulsion}} = 1/4.17 \times 10^{42}$$

$$= 1/4,170,000,000,000,000,000,000,000,000,000,000,000,000,000,000$$

After Feynman

We now turn to the kinetic energy. Here there is an important difference between the solar system and the atom. In the former, the kinetic energy depends very much on what the planet had to start with when the solar system was formed. The net result is that in the solar system, there are no such things as preferred orbits. In an atom, the story is different.

You might remember that one of Bohr's achievements was that he figured out for the first time that there existed preferred orbits. Not only that, he also gave a rule for specifying these orbits—this is the Bohr quantisation rule.

There is another way of stating this rule, that is by remembering that electrons can behave both as particles as well as waves, i.e. de Broglie waves. What we suppose is that there are de Broglie waves wrapped around each Bohr orbit as sketched in Fig. 3.1. Caution: The wavelength is not the same in the various orbits! The wavelength is determined by the kinetic energy, and the radius is determined by the balance between the potential and the kinetic energies.

The kinetic energy is given by

$$K = (1/2)m_e v^2 = (1/2)(m_e v)^2/m_e = p^2/2m_e \tag{3.2}$$

where $p = m_e v$ is the momentum of the electron. Now according to de Broglie, the electron wavelength $\lambda$ is related to momentum by

$$\lambda = h/p \tag{3.3}$$

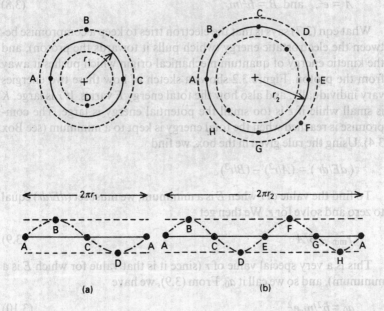

**Fig. 3.1** De Broglie waves wrapped around electron orbits. (a) shows the one-wavelength case, and (b) the two-wavelength case. Points A, B, C, ... etc., show the various crossings, maxima and minima. For viewing convenience, the waves have been unwrapped and shown below the orbits.

## Why are things the way they are?

If one wave is wrapped around the orbit, we also have

$$\lambda = 2\pi r \tag{3.4}$$

Combining the above two equations,

$$r = (\lambda/2\pi) = \hbar/p \tag{3.5}$$

Hence, kinetic energy $K$ is given by

$$K = p^2/2m_e = \hbar^2/2m_e r^2 \tag{3.6}$$

So the total energy $E$ is

$$E = -e^2/r + \hbar^2/2m_e r^2 \tag{3.7a}$$

or,

$$E = -A/r + B/2r^2 \tag{3.7b}$$

where

$$A = e^2, \text{ and } B = \hbar^2/m_e \tag{3.8}$$

What eqn (3.7) says is that the electron tries to keep a compromise between the electrostatic energy (which pulls it towards the proton), and the kinetic energy of quantum mechanical origin which pushes it away from the proton. Figure 3.2 shows a sketch of how these two energies vary individually, and also how the total energy $E$ varies. If $r$ is large, $K$ is small while if $r$ is too small the potential energy is large. The compromise is reached when the total energy is kept to a minimum (see Box 3.4). Using the rule given in the box, we find

$$(dE/dr) = (A/r^2) - (B/r^3)$$

To find the value of $r$ when $E$ is a minimum, we must set $(dE/dr)$ equal to zero and solve for $r$. We then get

$$r_{min} = B/A \tag{3.9}$$

This is a very special value of $r$ (since it is that value for which $E$ is a minimum), and so we call it $a_0$. From (3.9), we have

$$a_0 = \hbar^2/m_e e^2 \tag{3.10}$$

One thus gets

$$E_0 = -(m_e e^4/2\hbar^2) \tag{3.11}$$

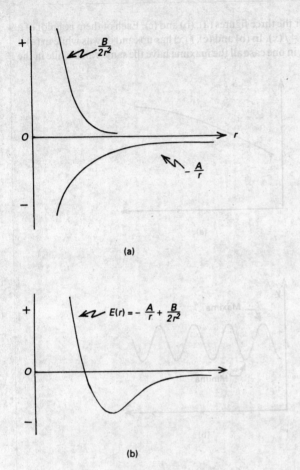

**Fig. 3.2** Schematic plot of $E = -(A/r) + (B/2r^2)$. (a) shows the two contributions separately, while (b) shows the composite. The net effect of the two terms together is to produce a minimum. Observe that for very small values of $r$, $E \sim B/2r^2$ while for very large $r$, $E \sim (-A/r)$.

If you substitute numerical values, you will find that

$$E_0 \sim 13.6 \text{ eV} = 1\,Ry$$

We call $a_0$ the *Bohr radius*.

Next let us look at the case where the electron is in the second Bohr orbit. The energy will be higher and so we call this an *excited state*. The state corresponding to the orbit with radius $a_0$ is called the *ground state*.

**Box 3.4** Consider the three figures (a), (b) and (c). Each of them is a plot of a certain function $y = f(x)$. In (b) and (c), $f(x)$ has ups and downs while in (a) it does not. Further, in one case all the maxima have the same value while in the

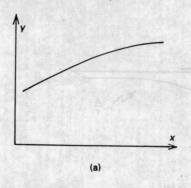

(a)

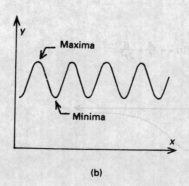

(b)

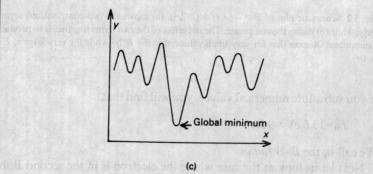

(c)

other, they have different values; the same is true of the minima. In (c) there is one minimum which is lower than the others; it is called the global minimum.

There is a simple rule for finding the values of $x$ for which $y$ is either a minimum or a maximum. The rule says: (i) Calculate $(dy/dx)$. (ii) Put $(dy/dx)$ equal to 0. (iii) Solve for $x$. The values obtained correspond to those for which $y$ is either a maximum or a minimum.

How do we know which is which? Simple. Calculate $(d^2y/dx^2)$ for those same values of $x$. If this quantity is –ve then $y$ has a maximum at the corresponding value of $x$; if it is +ve, then $y$ has a minimum. (Actually, a Pundit would caution and say: 'If $d^2y/dx^2$ is zero but $d^3y/dx^3$ is not, then $y$ would have an inflection at that point.')

Example:

$y = f(x) = \sin x$

$(dy/dx) = \cos x$

$\cos x = 0$ when $x = (2n+1)\pi/2$, $n = 0, \pm 1, \pm 2, \ldots$

So $y$ has maxima or minima at these values of $x$.

$(d^2y/dx^2) = -\sin x$

This quantity is –ve at $x = \pm\dfrac{\pi}{2}, \pm\dfrac{5\pi}{2}, \pm\dfrac{9\pi}{2}, \pm\dfrac{13\pi}{2}, \ldots$

So $y$ has a maximum at these values of $x$.

$(d^2y/dx^2)$ is +ve at $x = \pm\dfrac{3\pi}{2}, \pm\dfrac{7\pi}{2}, \pm\dfrac{11\pi}{2}, \pm\dfrac{11\pi}{2}, \ldots$

So $y$ has a minimum at these values of $x$.

What we now want is the radius of the orbit corresponding to the first excited state. The first thing to note is that two wavelengths would be wrapped around the orbit in this case, i.e. $2\pi r = 2\lambda$ or $r = \lambda/\pi$. So,

$$K = (p^2/2m_e) = (h^2/2m_e\lambda^2)$$
$$= 4(\hbar^2/2m_e r^2) \qquad (3.12)$$

Potential energy is given as before by eqn (3.1b). Once again we add the two energies to obtain an expression for the total energy $E$, and then find the radius of the orbit by finding the value of $r$ for which the energy is a minimum. The steps are the same as before, and so I shall not repeat them. The final result is that

$$r_1 = 4a_0; \text{ also, } E_1 = E_0/4 = -0.25\,Ry \qquad (3.13)$$

We find that the radius of the second Bohr orbit is four times larger and so is the energy (remember, $-0.25$ is greater than $-1.0$). One can use similar arguments to show that the radius of the $n$th orbit is $n^2 a_0$, while the energy is $(1/n^2)E_0$.

You might be wondering if the back-of-the-envelope business is a joke, since I have taken a few pages to do my derivations. My discussion was long because I was explaining the various steps in detail. A professional physicist would compress all these things into a few steps, making the calculation truly of the back-of-the-envelope type; see Box 3.5.

---

**Box 3.5** Back-of-the-envelope estimate of the Bohr radius and the energy in the first Bohr orbit.

$$\text{Kinetic energy} = (\hbar^2/2m_e r^2); \text{ potential energy} = -(e^2/r)$$
$$\text{Total energy} = (\hbar^2/2m_e r^2) + (-e^2/r)$$
$$= -(A/r) + (B/r^2)$$
$$(dE/dr) = (A/r^2) - (B/r^3) = 0$$
$$\text{implies } r_{min} = a_0 = (B/A) = \hbar^2/m_e e^2.$$
$$E_0 = \text{Energy in orbit of radius } a_0$$
$$= -(A/a_0) + (B/2a_0^2)$$
$$= -(m_e e^4/2\hbar^2)$$

Maybe the envelope would appear a bit crowded, but the result can surely be squeezed in!

---

## 3.2 The helium atom

We now play the same game for the helium atom; but there is a complication here, which is why we are considering the case specially. As you are aware, there are two electrons in the helium atom. In the ground state, both these are in the same orbit circling around the nucleus. Once again we have to do the balancing act, but there is now a new element in the picture, namely the electrostatic repulsion between the two electrons. For a moment, let us ignore this repulsion. Proceeding as before, we have for the potential energy,

$$V = (-2e^2/r) + (-2e^2/r)$$
$$= -4(e^2/r) \tag{3.14}$$

Likewise,

$$K = 2(\hbar^2/2m_e r^2) \tag{3.15}$$

So the total energy is given by

$$E = -4(e^2/r) + 2(\hbar^2/2m_e r^2) \quad (3.16)$$

From this we can derive as before the radius and the energy, and we find

$$r = (2a_0/4) = 0.5 a_0 \text{ and}$$
$$E = 16E_0/2 = -8\,Ry \quad (3.17)$$

These results are certainly not correct, since we have neglected the repulsion between the two electrons; time to bring that in. We can calculate the repulsion if we know how far apart the two electrons are. This we don't know but we can make a guess (here is a new word for you; guesstimate = guess + estimate!). Look at Fig. 3.3. The maximum distance possible between the two electrons is clearly $2r$. The minimum distance is of course zero but that would lead to infinite energy which is unacceptable.

Let us see what Weisskopf does. He says that *on the average* the two electrons are at a distance $r_{eff}$ apart and further that

$$r < r_{eff} < 2r$$

Weisskopf now just assumes

$$r_{eff} = r/0.6 \quad (3.18)$$

This is undoubtedly a guess, but nevertheless a reasonable one. You could as well have assumed

$$r_{eff} = r/0.8$$

and that would be just as good, given the spirit of the game. What is more important is to appreciate how one manoeuvres. Notice the remarks in the caption of Fig. 3.3.

OK. So we have an estimate for $r_{eff}$. Using this, the repulsive energy is given by

$$+(e^2/r_{eff}) = 0.6(e^2/r) \quad (3.19)$$

Therefore the total energy is

$$E = (e^2/r)[-4 + 0.6] + 2(\hbar^2/2m_e r^2) \quad (3.20)$$

As before we minimise $E$ to find that the radius of the orbit is given by

$$r = (2/3.4)a_0 \sim 0.6 a_0 \quad (3.21)$$

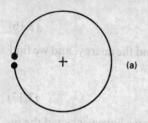

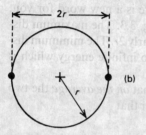

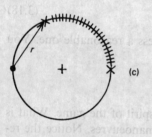

**Fig. 3.3** Disposition of the two electrons in the lowest orbit of helium. (a) illustrates one extreme case which is ruled out because of infinite repulsion energy. (b) shows the maximum distance apart the electrons can be. Weisskopf makes a guess that the second electron is somewhere in the shaded patch in (c). Actually, if we simply choose two points at random on the circumference, the average distance between them is approximately $r/0.8$. Weisskopf chooses the value $r/0.6$ in order to be close to the correct answer given in (3.22)!

and the energy by

$$E = -5.8 \, Ry \tag{3.22}$$

which, incidentally, is close to the experimental value.

There is a small lesson one can draw from all this; in fact, we shall be using that lesson shortly. Look back at eqn (3.21). We see that the radius of the first or the innermost orbit is

$$r \sim 0.5 \, a_0 = (1/A) a_0 \tag{3.23}$$

where $A$ is the atomic number of helium ($=2$). In general, this is true, i.e. the radius of the *first* orbit in any atom is approximately $(1/A)a_0$, where $A$ is the atomic number of the atom.

## 3.3 The neon atom

Neon has 10 electrons, arranged in two groups or shells as shown in Fig. 3.4. If you want to know more about electronic shells, see Box 3.6. Now neon is quite a complicated atom. Let us see how Weisskopf handles it. He is interested only in the L-shell radius, since the estimate for the K-shell can be had from the formula (3.23).

Now what do the electrons in the L-shell of neon see? They see the nucleus which has a charge $+10e$. But they also see the two electrons in the K-shell. So the *net* charge they see is $(10-2)e = 8e$. In other words, we shall calculate the L-shell radius somewhat like in the helium problem, but assuming an effective nuclear charge $z_{eff} = +8e$.

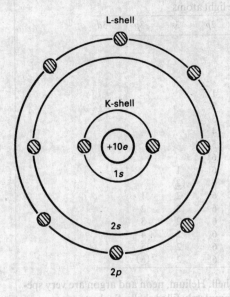

| 1s | 2 | closed shell | — K |
| 2s | 2 | closed shell } | |
| 2p | 6 | closed shell } | — L |

**Fig. 3.4** Schematic drawing of the neon atom showing the organisation of the electrons into shells.

It is easy to write down the expressions for the potential energy $V$ and the kinetic energy $K$. We have

$$V = -(8e)^2/r \qquad (3.24)$$

To calculate the kinetic energy $K$, we suppose that the L-shell can be looked upon as the second Bohr orbit of an atom with nuclear charge $8e$.

**Box 3.6** In the simple Bohr model, we have just one electron 'orbiting' around the nucleus. In reality, atoms contain many electrons—in fact, in the neutral atom there are as many electrons as there are protons in the nucleus. The question now is: 'How are these electrons organised?' And the simple answer is: 'In shells.' The shells can be grouped into broad categories called the first shell (also known as the K-shell), the second (or the L) shell and so on. They are given numbers 1,2,3... etc. Other than the K-shell, all the other shells have subshells. Thus the second shell has the subshells $2s$ and $2p$, the third one has $3s$, $3p$, $3d$ and so forth. There is a limit to the maximum number of electrons one can put in any one of these subshells—in the $s$-type it is 2, in the $p$-type it is 6, in the $d$-type it is 10, etc. Based on this, the electronic configuration of the first few elements of the periodic table is as below:

Electronic configurations of some light atoms

| Z  | Element | 1s  | 2s  | 2p  | 3s  | 3p  | 3d  |
|----|---------|-----|-----|-----|-----|-----|-----|
| 1  | H       | 1   |     |     |     |     |     |
| 2  | He      | (2) |     |     |     |     |     |
| 3  | Li      | 2   | 1   |     |     |     |     |
| 4  | Be      | 2   | (2) |     |     |     |     |
| 5  | B       | 2   | 2   | 1   |     |     |     |
| 6  | C       | 2   | 2   | 2   |     |     |     |
| 7  | N       | 2   | 2   | 3   |     |     |     |
| 8  | O       | 2   | 2   | 4   |     |     |     |
| 9  | F       | 2   | 2   | 5   |     |     |     |
| 10 | Ne      | 2   | 2   | (6) |     |     |     |
| 11 | Na      | 2   | 2   | 6   | 1   |     |     |
| 12 | Mg      | 2   | 2   | 6   | (2) |     |     |
| 13 | Al      | 2   | 2   | 6   | 2   | 1   |     |
| 14 | Si      | 2   | 2   | 6   | 2   | 2   |     |
| 15 | P       | 2   | 2   | 6   | 2   | 3   |     |
| 16 | S       | 2   | 2   | 6   | 2   | 4   |     |
| 17 | Cl      | 2   | 2   | 6   | 2   | 5   |     |
| 18 | Ar      | 2   | 2   | 6   | 2   | (6) |     |

Here a circle represents a closed shell. Helium, neon and argon are very special atoms because they all have completely filled shells. Such atoms are very reserved, and do not generally like either to receive or donate electrons. Atoms like Li and Na on the other hand, are just the opposite. In chemistry, we study how atoms combine to form molecules or how molecules break up. All these are possible only when the electrons in the outer shells of atoms can jump around. For further details about the electronic structure of the atoms, see the periodic table in Box 3.7.

Thus for every electron in this shell there is a contribution of $4(\hbar^2/2m_e r^2)$ — see (3.12); so,

$$K = 8 \times 4(\hbar^2/2m_e r^2) \qquad (3.25)$$

## How big is an atom?

As in the helium problem, the total energy $E$ is not given simply by $(V + K)$; there is also the repulsion energy to be considered. Really speaking, the calculation of this repulsive energy is no mean task. But see how skilfully Weisskopf handles it. He says: 'Let us form pairs out of the electrons in the L-shell. There are 8 electrons, and so we can form $(8 \times 7)/2 = 28$ distinct pairs.' I hope you do not find it difficult to understand how this number 28 is arrived at.

OK. So we have these 28 pairs. We now treat each pair as in the helium problem, i.e. we assume that the electrons belonging to a pair are at a distance $r_{eff}$ apart. Further, as in the helium problem, it is assumed that

$$r_{eff} = r/0.6$$

---

**Box 3.7** Collected together here are certain facts which supplement Table 3.1. First we have the periodic table. This shows how the electrons are arranged in shells.

Next comes the radii of atoms. Now atoms are not well-defined spheres like billiard balls; so when one talks of a radius, one has to be careful. Nevertheless, a meaningful radius can and has been defined, and values found from experimental data. Figure (a) shows a plot. From it we see that

1. the alkali elements Li, Na, ... etc., have large radii,
2. following each alkali element, there is a progressive shrinkage as one moves across a period and,
3. in a period, the closed shell atom has the smallest radius, e.g. neon.

The first ionization potential (f.i.p.) is the energy required to remove an electron from the outermost orbit. Figure (b) shows a plot of this quantity. As expected, it is high for the closed shell atoms.

What is the connection between the f.i.p. and the binding energy calculated in Table 3.1? In the case of lithium, they are one and the same. (Remember the units are different!) However, when there is more than one electron in the outer shell there is a problem; but this is to be expected. Take beryllium for example. To remove the first electron, an energy equal to f.i.p. must of course be supplied. The second electron will not budge with the same amount of energy. This is because with the removal of one electron, the neutral beryllium atom becomes a positive ion. The second electron must now be torn off a positive ion and this requires more energy than in the case of a neutral atom. In short,

$$2 \times \text{f.i.p.} < 2 \times E \text{ (where } E \text{ is the binding energy/electron)}$$
$$= (\text{f.i.p.}) + (\text{s.i.p.}).$$

In other words, the second ionization potential (s.i.p.) is greater than the first. If that difference is properly taken into account, then there would be no problem.

Box 3.7 (continued)

## Periodic Table, with the Outer Electron Configurations of Neutral Atoms in Their Ground States

The notation used to describe the electronic configuration of atoms and ions is discussed in all textbooks of introductory atomic physics. The letters $s, p, d \ldots$ signify electrons having orbital angular momentum $0, 1, 2, \ldots$ in units $\hbar$; the number to the left of the letter denotes the principal quantum number of one orbit, and the superscript to the right denotes the number of electrons in the orbit.

| 1 | 2 | | | | | | | | | | | 13 | 14 | 15 | 16 | 17 | 18 |
|---|---|---|---|---|---|---|---|---|---|---|---|---|---|---|---|---|---|
| **H** 1 $1s$ | | | | | | | | | | | | | | | | | **He** 2 $1s^2$ |
| **Li** 3 $2s$ | **Be** 4 $2s^2$ | | | | | | | | | | | **B** 5 $2s^2 2p$ | **C** 6 $2s^2 2p^2$ | **N** 7 $2s^2 2p^3$ | **O** 8 $2s^2 2p^4$ | **F** 9 $2s^2 2p^5$ | **Ne** 10 $2s^2 2p^6$ |
| **Na** 11 $3s$ | **Mg** 12 $3s^2$ | | | | | | | | | | | **Al** 13 $3s^2 3p$ | **Si** 14 $3s^2 3p^2$ | **P** 15 $3s^2 3p^3$ | **S** 16 $3s^2 3p^4$ | **Cl** 17 $3s^2 3p^5$ | **Ar** 18 $3s^2 3p^6$ |
| **K** 19 $4s$ | **Ca** 20 $4s^2$ | **Sc** 21 $3d$ $4s^2$ | **Ti** 22 $3d^2$ $4s^2$ | **V** 23 $3d^3$ $4s^2$ | **Cr** 24 $3d^5$ $4s$ | **Mn** 25 $3d^5$ $4s^2$ | **Fe** 26 $3d^6$ $4s^2$ | **Co** 27 $3d^7$ $4s^2$ | **Ni** 28 $3d^8$ $4s^2$ | **Cu** 29 $3d^{10}$ $4s$ | **Zn** 30 $3d^{10}$ $4s^2$ | **Ga** 31 $4s^2 4p$ | **Ge** 32 $4s^2 4p^2$ | **As** 33 $4s^2 4p^3$ | **Se** 34 $4s^2 4p^4$ | **Br** 35 $4s^2 4p^5$ | **Kr** 36 $4s^2 4p^6$ |
| **Rb** 37 $5s$ | **Sr** 38 $5s^2$ | **Y** 39 $4d$ $5s^2$ | **Zr** 40 $4d^2$ $5s^2$ | **Nb** 41 $4d^4$ $5s$ | **Mo** 42 $4d^5$ $5s$ | **Tc** 43 $4d^6$ $5s$ | **Ru** 44 $4d^7$ $5s$ | **Rh** 45 $4d^8$ $5s$ | **Pd** 46 $4d^{10}$ | **Ag** 47 $4d^{10}$ $5s$ | **Cd** 48 $4d^{10}$ $5s^2$ | **In** 49 $5s^2 5p$ | **Sn** 50 $5s^2 5p^2$ | **Sb** 51 $5s^2 5p^3$ | **Te** 52 $5s^2 5p^4$ | **I** 53 $5s^2 5p^5$ | **Xe** 54 $5s^2 5p^6$ |
| **Cs** 55 $6s$ | **Ba** 56 $6s^2$ | **La** 57 $5d$ $6s^2$ | **Hf** 72 $4f^{14}$ $5d^2$ $6s^2$ | **Ta** 73 $5d^3$ $6s^2$ | **W** 74 $5d^4$ $6s^2$ | **Re** 75 $5d^5$ $6s^2$ | **Os** 76 $5d^6$ $6s^2$ | **Ir** 77 $5d^9$ — | **Pt** 78 $5d^9$ $6s$ | **Au** 79 $5d^{10}$ $6s$ | **Hg** 80 $5d^{10}$ $6s^2$ | **Tl** 81 $6s^2 6p$ | **Pb** 82 $6s^2 6p^2$ | **Bi** 83 $6s^2 6p^3$ | **Po** 84 $6s^2 6p^4$ | **At** 85 $6s^2 6p^5$ | **Rn** 86 $6s^2 6p^6$ |
| **Fr** 87 $7s$ | **Ra** 88 $7s^2$ | **Ac** 89 $6d$ $7s^2$ | | | | | | | | | | | | | | | |

### Lanthanides

| **Ce** 58 $4f^2$ $6s^2$ | **Pr** 59 $4f^3$ $6s^2$ | **Nd** 60 $4f^4$ $6s^2$ | **Pm** 61 $4f^5$ $6s^2$ | **Sm** 62 $4f^6$ $6s^2$ | **Eu** 63 $4f^7$ $6s^2$ | **Gd** 64 $4f^7$ $5d$ $6s^2$ | **Tb** 65 $4f^8$ $5d$ $6s^2$ | **Dy** 66 $4f^{10}$ $6s^2$ | **Ho** 67 $4f^{11}$ $6s^2$ | **Er** 68 $4f^{12}$ $6s^2$ | **Tm** 69 $4f^{13}$ $6s^2$ | **Yb** 70 $4f^{14}$ $6s^2$ | **Lu** 71 $4f^{14}$ $5d$ $6s^2$ |
|---|---|---|---|---|---|---|---|---|---|---|---|---|---|

### Actinides

| **Th** 90 — $6d^2$ $7s^2$ | **Pa** 91 $5f^2$ $6d$ $7s^2$ | **U** 92 $5f^3$ $6d$ $7s^2$ | **Np** 93 $5f^4$ $6d$ $7s^2$ | **Pu** 94 $5f^6$ $7s^2$ | **Am** 95 $5f^7$ $7s^2$ | **Cm** 96 $5f^7$ $6d$ $7s^2$ | **Bk** 97 | **Cf** 98 | **Es** 99 | **Fm** 100 | **Md** 101 | **No** 102 | **Lr** 103 $4f^{14}$ $5d$ $6s^2$ |
|---|---|---|---|---|---|---|---|---|---|---|---|---|---|

How big is an atom 41

Box 3.7 (*continued*)

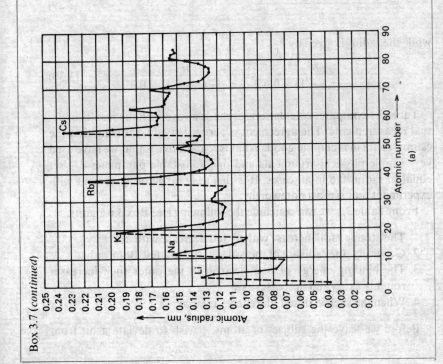

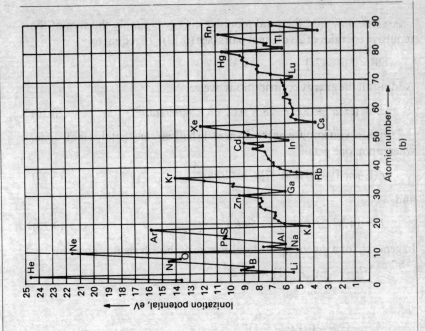

Assume now that the total repulsive energy is simply the sum of the repulsive energies of 28 distinct pairs (see (3.19)), i.e. equal to

$$28 \times 0.6(e^2/2r)$$

Adding all the energy contributions, we get

$$E = (e^2/2r)[-(8)^2 + 28 \times 0.6] + 8 \times 4(\hbar^2/2m_e r^2) \qquad (3.26)$$

I suppose you know what to do next. I shall assume that this is all done, whereupon one gets

$$r(\text{L-shell}) = [(8 \times 4)/47]a_0 \sim (2/3)a_0 \qquad (3.27)$$

and

$$E = (47)^2 E_0/(8 \times 4) \sim -69\, Ry \qquad (3.28)$$

In general, for a given $z_{\text{eff}}$, the total energy of the $n$th shell is given by

$$E = \frac{[z_{\text{eff}}^2 - z_{\text{eff}}(z_{\text{eff}} - 1) \times 0.6/2]^2}{n^2\, z_{\text{eff}}}$$

$$= \frac{z_{\text{eff}}[z_{\text{eff}} - 0.3(z_{\text{eff}} - 1)]^2}{n^2} \qquad (3.29)$$

while the radius is given by

$$r = \frac{z_{\text{eff}}\, n^2}{z_{\text{eff}}^2 - z_{\text{eff}}(z_{\text{eff}} - 1) \times 0.6/2} = \frac{n^2}{z_{\text{eff}} - (z_{\text{eff}} - 1) \times 0.3} \qquad (3.30)$$

Table 3.1 compares calculations as made above with the results of actual measurements. The agreement is very satisfying, showing that the estimates are in fact quite good. What is remarkable is that we have used very simple physical arguments and one adjustable parameter $r_{\text{eff}}$, set equal to radius/0.6 in all cases. It is this *wide* range of agreement with experiment that is most noteworthy.

From Table 3.1 we can conclude the following (see Box 3.6 again):

1. The atomic radii increase with $n^2$ and decrease with $z$.
2. Generally, the radii are of the order of 0.5 to a few Bohr radii.
3. The binding energy of one electron in the outer shell increases roughly as $z_{\text{eff}}^2$ and decreases as $n^2$.
4. When the shell is closed, the energy is large.

Before we leave the subject of atoms, I wish to deviate a bit from

## Table 3.1

| Element | z | $z_{eff}$ | n | r calculated | E calculated | r measured | E measured |
|---|---|---|---|---|---|---|---|
| H  | 1  | 1 | 1 | 1.0 | 1.0  | 1.0 | 1.0  |
| He | 2  | 2 | 1 | 0.6 | 5.8  | 0.6 | 5.8  |
| Li | 3  | 1 | 2 | 4.0 | 0.25 | 2.8 | 0.4  |
| Be | 4  | 2 | 2 | 2.4 | 1.4  | 2.2 | 2.0  |
| B  | 5  | 3 | 2 | 1.7 | 4.3  | 1.6 | 5.2  |
| C  | 6  | 4 | 2 | 1.3 | 9.6  | 1.2 | 10.9 |
| N  | 7  | 5 | 2 | 1.1 | 18.0 | 1.0 | 19.5 |
| O  | 8  | 6 | 2 | 0.9 | 30.5 | 0.8 | 31.8 |
| F  | 9  | 7 | 2 | 0.8 | 42.0 | 0.7 | 48.5 |
| Ne | 10 | 8 | 2 | 0.7 | 69.0 | 0.6 | 70.0 |

Note: Radii are given in units of $a_0$ and energy in units of $Ry$.

Weisskopf's lectures to refer to a lecture given some years ago by the famous astrophysicist Chandrasekhar (see Box 3.8) in Ahmedabad. Chandrasekhar was lecturing about Black Holes (we shall meet them again in Chapter 7) but he began his lecture with a brief reference to atoms just to illustrate how fundamental constants exert their influence. He asked: 'Why are the atoms as they are?' By way of answering this question, he then gave a brief description of Bohr's atom model. He

---

**Box 3.8** Subrahmanyam Chandrasekhar is perhaps the greatest mathematical physicist India has produced. He was born in Lahore (now in Pakistan and then in undivided British India) on 19 October 1910 (19/10/1910, easy to remember!) His father C.Subrahmanya Iyer (popularly known as C.S.Iyer) was the elder brother of C.V.Raman. Like Raman, Chandrasekhar studied in the Presidency College, Madras, and like Raman he too won the Nobel Prize for Physics. So that makes two in the family, both students of Presidency. It would be a long time before we see another college in India beat that record.

It would need a whole book to do justice to the fabulous contributions which Chandrasekhar has made and is still making. Here let me just give you a flavour of his style. Firstly, he is very careful and precise in his statements—both oral and written. Secondly, he is always very clear—no fuzzy nonsense with him. Thirdly, he is always very thorough—no loose ends ever. Fourthly, and this also is very important, he is *incredibly* serious about his work.

His style is to choose a subject, work on it for a few years, and then write a book on it. After that there is usually very little left for the others to do. As someone said, ' Chandra always seems to have the last word on the subject.' Every book of his is a classic and contains new mathematical methods which

> are widely used. Do you know that after he turned seventy, he took up a new subject—*The Mathematical Theory of Black Holes*, and as usual wrote a masterpiece. Not easy to beat that. I know many who lead a retired life at half that age!
>
> One can go on and on about Chandrasekhar. If you want a hero in physics, here is one. They don't come any better. See also the companion volume *Chandrasekhar and His Limit*.

then asked: 'Why are there just 92 naturally occurring elements? Why are there not a thousand or ten thousand different atomic species?' Here I must interrupt to say that new elements have been made in the laboratory and in this manner the periodic table *has been extended* to about element number 105. But the point Chandrasekhar is making is that we don't seem to have elements with atomic numbers like 856, for example. Why? This is what he says:

> The electron around the hydrogen nucleus circulates with a certain velocity, say, $v$. Now according to the Bohr model, the radius of a singly charged helium atom will be one-fourth the radius of the hydrogen atom, and the electron will circulate around the nucleus with a velocity which is twice as large. Thus, as the nuclear charge increases, the electron orbits come closer to the centre and the velocity of circulation becomes larger; and eventually the velocity of the electron will approach the velocity of light. But according to the theory of special relativity, no particle can have a velocity exceeding that of light. This is signalled by the fact that when particles move with a velocity close to that of light, their effective masses increase indefinitely. On this account, the atomic theory must be modified to take into account the effects predicted by the theory of special relativity. And by including the effects of special relativity, one can show that there is a maximum charge for which an electron circulating around it can be stable. This is given by the reciprocal of the so-called fine structure constant, namely, $hc/(2\pi e^2) \sim 137$. The theory, therefore, shows that one can have no stable atomic structure with a nuclear charge in excess of 137. Of course, the number of naturally occurring stable elements is 92, which has been increased beyond 100 by artificial methods. Nevertheless, that we have only around 100 different atomic species is determined by the combination of the natural constants, $hc/(2\pi e^2)$.

You see here another beautiful example of quick reasoning based on simple arguments and the fundamental constants. The big shots do this all the time.

# 4 On Solids And Liquids

We shall now look at a few properties of solids and liquids. In the past, solids and liquids were often treated as separate entities, leading to solid state physics on the one hand and liquid state physics on the other. Nowadays, the two are treated together under the name condensed matter physics.

As you know, even in ancient times people asked what happens if one keeps on dividing a solid into smaller and smaller portions; and in this way they arrived at the concept of atoms. We are now about to play a reverse game, that is, knowing the properties of atoms, can we predict the properties of the condensed state? This is the essence of condensed matter physics—to explain *macroscopic* properties in terms of *microscopic* physics. Of course if one took this task very seriously, then one would have quite a big problem on hand. We are therefore going to be content with making a few estimates by our usual method.

Let us start by recalling what we know about atoms—at least what we need to know for the present purpose. We already know that the electrons in an atom are arranged in shells; we have in fact calculated their radii and their energies. We know that the way atoms combine depends very much on how many electrons there are in the outermost shells; this infact is the basis of modern chemistry (see Box 4.1).

---

**Box 4.1** We have seen earlier (Box 3.6) how electrons are distributed amongst the various shells. We now consider the bonding of atoms to form molecules and solids.

Atoms join together to form molecules. So some cement or bridge is needed to keep the atoms together. For example, in the water molecule, we have two atoms of H and one of O all linking together. What holds them? Answer: The bonds formed by the outer electrons in the various atoms. Chemistry is largely the story of these bonds. A glimpse now into the rules of the game.

Consider sodium which has one electron in the $3s$ shell. This electron is eager to cut loose and does so whenever another atom like chlorine, which is greedy for an electron, comes along. The $3p$ shell of chlorine is not full, and it is willing to receive a guest. So sodium and chlorine readily team up, forming an *ionic* bond. Thus one of the rules of bond formation is that swapping must leave the participating atoms with closed-shell configurations. This clearly happens in the formation of the NaCl bond as illustrated in figure (a).

Another important type of bond is the covalent bond. The hydrogen molecule provides an example. This molecule is formed by two atoms, each of which tries to grab the other's electron; a compromise is then reached and the electron pair is shared by both the protons—see (b). Covalent bonds form an important cornerstone of chemistry.

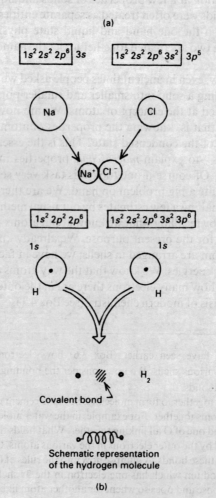

Schematic representation of the hydrogen molecule

(b)

Usually there is only a small number of atoms in a molecule. True, in molecules like the proteins and DNA, the number is quite large; but even so, the number is nothing like what it is in a solid or liquid where one has to deal with an Avogadro number of atoms.

OK. So a solid is made by bringing together something like $10^{23}$ atoms. When they are brought together, they stay together; the atoms don't fly apart. Why? Because the energy of atoms packed together is lower than the energy of separated atoms (Fig. 4.1). For obvious reasons,

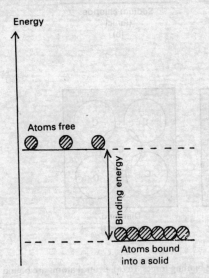

Fig. 4.1 Atoms bind together and stay as a solid because the energy of that configuration is lower than that in which the atoms are neutral, free and infinitely far apart. The difference in energy is called the *binding energy*. If a heat source were to supply to a solid an energy of the order of binding energy, it would encourage the atoms to break apart and become free. As we shall see, a solid can 'come apart' even with a lesser energy input.

this energy is called the *binding* energy. We would now like to estimate this binding energy, but before we do that let us briefly discuss the different ways in which solids can be formed, i.e. the different types of solid-state binding.

The formation of solids is governed to some extent by the same rules which determine the formation of molecules. Based on this, one can identify four important types of solids, as illustrated in Fig. 4.2.

To discuss cohesion or binding, we compare the total energy of the solid, i.e. the kinetic energy of all the atoms plus the potential energy, with the energy of the same number of *free neutral atoms at infinite separation*. However, when we make our estimates, we shall assume that the solid is at zero temperature (i.e. 0 °K ). Then the atoms would all be absolutely still and therefore have no kinetic energy. So we can get rid of the kinetic energy term, and not bother about calculating it.

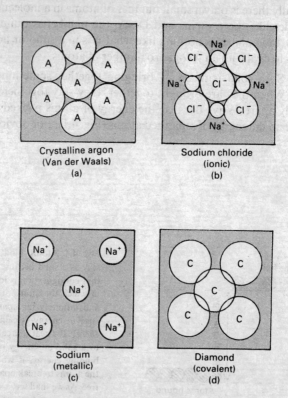

**Fig. 4.2** The principal types of crystalline binding forces. In (a), neutral atoms are bound together by weak forces called van der Waals forces. Solid argon and solid neon are examples. Observe that both argon and neon are closed-shell atoms. In crystals like anthracene, one has a somewhat similar packing of neutral molecules. In (b), the electrons are transferred, leading to strong, attractive electrostatic forces which provide binding. Metals (see c) are formed when valence electrons cut loose from their respective parent atoms and form an all-pervading 'electron gas' which acts like a cement or glue. Covalent-bonded solids like covalent-bonded molecules are also known, the classic example being diamond—see (d). Here the bond is so strong as to make diamond the hardest known substance. The covalent bond is described in Box 4.1.

## 4.1 Binding energy of ionic solids

Now, let us see how Weisskopf calculates the binding energy of ionic solids. Why ionic solids and not any other type of solid? Well, there is a good reason. Let us suppose the atoms to be arranged in a lattice, and that we can inflate or deflate the lattice as we please (Fig. 4.3). We shall now plot the (potential) energy $E(R)$ of the lattice as a function of the

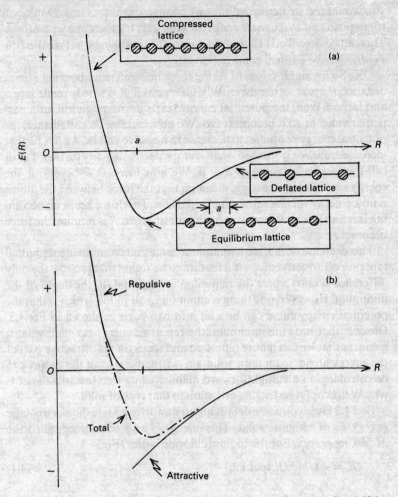

**Fig. 4.3** Imagine a crystalline solid which can be uniformly inflated or deflated. If the energy $E(R)$ of the lattice is plotted as a function of the lattice spacing $R$, one would obtain a curve as sketched in (a). Obviously, the lattice would prefer the spacing $R = a$. $E(R)$ is a composite curve made up of a short-range part that represents repulsion and a long-range part that represents attraction as in (b). At $R = a$, $E(R)$ is dominated by the attractive part. In an ionic solid, the attractive part is almost entirely elctrostatic in origin.

lattice constant $R$ (Fig. 4.3). When $R$ is very small, the energy will be very large and positive. Moreover, $E(R)$ decreases as $R$ is increased. This means that the force between the atoms is repulsive (recall from Box 3.3, force $= -dV/dr$). There is repulsion for small $R$ values because the atoms are very crowded together and they don't like it one bit. Imagine how you would feel in a crowded railway compartment or a bus. Don't you try to push and make some room for yourself? It's the same with atoms. This can be described in a more technical fashion using what is called the *Pauli principle*—I shall come to that later.

OK. So for small values of $R$, the force between neighbouring atoms, distance $R$ apart, is repulsive. What happens if $R$ is slowly made larger and larger? Well, the potential energy starts decreasing until, at a particular value of $R$, it becomes zero. We now increase $R$ still further; the potential energy continues to decrease to negative values. As $R$ is further increased, a stage is reached when it stops decreasing any further. Let us call this particular value of $R$ as $a$. We now increase $R$ beyond $a$; the energy now starts increasing, meaning that the force between the atoms is now a net *attraction* rather than repulsion. This force keeps on becoming less and less as $R$ increases further until, when $R$ is infinite, the force becomes zero.

This decrease of $E(R)$ to a minimum value and its subsequent gradual tapering off to zero energy is a feature to be found in all solids; the only difference is in (i) where the minimum occurs, and (ii) the depth of the minimum. However, one thing is common to all solids, which is that the potential energy curve can be split into two parts as shown in Fig 4.3. Observe that near the minimum, the repulsive energy is small; what it means is that we can ignore repulsion and focus on the attractive part of the energy alone. In an ionic solid, this attractive part of the energy can be calculated in a straightforward manner using electrostatics. That is why Weisskopf is restricting attention to that type of solid.

Box 4.2 gives you some details about how to calculate the electrostatic energy $E_c$ of an ionic solid. This energy is also sometimes called the *Madelung energy*. For the sodium chloride lattice (Fig. 4.4)

$$E_c = -1.74e^2/R \text{ (per ion)} \tag{4.1}$$

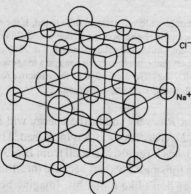

**Fig. 4.4** Schematic representation of the sodium chloride lattice. The crystal structure may be constructed by arranging $Na^+$ and $Cl^-$ ions alternately at the lattice points of a simple cubic lattice. In the crystal, each ion is surrounded by six nearest neighbour ions.

**Box 4.2** Consider an ionic lattice as in the figure. The lattice spacing is $R$ and all charges have the same *magnitude*, namely, $q$.

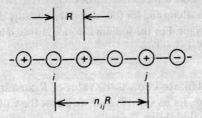

Consider two ions $i$ and $j$. The electrostatic interaction between them (i.e. the potential energy of the pair) is

$$U_{ij} = (\pm q^2/r_{ij})$$

where $r_{ij}$ is the distance between them. The + sign is taken if the charges at the two sites are *like*, and the − sign if the charges are *unlike*. For convenience, we write $r_{ij}$ as equal to $n_{ij}R$, where $n_{ij}$ is an integer.

We now ask for the energy of interaction of ion $i$ with all the other ions in the lattice. This is obviously equal to

$$\sum_j' U_{ij}$$

where $\sum'$ denotes a sum over $j$ excluding $j = i$. Using the formula for $U_{ij}$,

$$\sum_j' U_{ij} = \sum_j' q^2 (\pm / n_{ij} R)$$

If we write

$$\sum_j' U_{ij} = -(\alpha q^2/R),$$

then clearly

$$\alpha = -\sum_j' (\pm/n_{ij})$$

Let us now evaluate this sum for the chain in the figure, taking a −ve ion as the reference ion $i$. We then have

$$(\alpha/R) = [(2/R) - (2/2R) + (2/3R) - (2/4R)\ldots]$$
$$= (2/R)[1 - (1/2) + (1/3) - (1/4)\ldots]$$

Now

$$\log_e (1+x) = x - (x^2/2) + (x^3/3) - (x^4/4)\ldots$$

Using this we then get

> $\alpha = 2 \log_e 2$
>
> This shows how to calculate $\alpha$ in one dimension. In three dimensions the calculation is a bit more difficult but can be done. In fact, it was first done a long time ago by Madelung and for this reason the quantity $\alpha$ is referred to as the Madelung constant. For the sodium chloride lattice, it has the value ~ 1.74, which is what Weisskopf uses.

What we would like to know is the value of $R$ corresponding to the minimum. As before, let us say it is $a$. If we know the value of $a$, then we can substitute it in the above formula and get an estimate of the binding energy, for it would be approximately equal to $E_c$.

Now we can think of $a$ as the sum of the radii of the positive and the negative ions. The *average* radii of ions are known from various other sources. For example, the average radius for the $Na^+$ ion is 0.88 Å and for the $Cl^-$ ion is 1.48 Å. We might therefore think that for NaCl, $a = (0.88 + 1.48)$ Å. Not quite. According to Weisskopf, if $r_1$ and $r_2$ denote the radii of the two ions, then

$$a = f(r_1 + r_2) \sim f \times 2a_0$$

where $a_0$ is the Bohr radius (recall eqn (3.10)), and $f$ is what Weisskopf calls a fudge factor for which he assumes a value in the range $1 < f < 4$. So, the potential energy per atom, which is also the same thing as the binding energy $B$ per atom, in the solid is:

$$B \sim E_c = -1.74 \, e^2/a = -0.87 e^2/fa_0 \tag{4.2}$$

What finally emerges from all this is that the binding energy (per atom) is of the order of several electron volts. If this amount of energy is supplied, the solid will break up. Remember that we have assumed the temperature to be zero. If the solid is at room temperature, we will obviously need less energy to break it up. I suppose you can understand why.

## 4.2 Lattice vibrations

We have just seen that solids are formed when it is energetically favourable for them to do so. But we also know that solids melt to become liquids, and liquids evaporate when heated. These things happen because when we supply heat, the atoms start vibrating more and more vigorously. In the process, they jiggle the interatomic bonds more and more violently. Melting and evaporation are connected with the disruption of the bonds due to this jiggling process. So, if a solid has to remain a solid, the jiggling should not be much. In other words, the *amplitude* of vibration should be much less compared to the bond length. We now wish to convince ourselves that this is so by making an estimate.

The argument goes as follows: We start by supposing that each atom is a harmonic oscillator (see Box 4.3), i.e. it executes oscillations like a simple pendulum does. In the case of a pendulum, the higher the energy, the greater is the amplitude; so also with our atomic oscillator. There is a small problem though with our little oscillator. In the case of a pendulum, it is possible to think of it as having a definite energy (which of

**Box 4.3** The harmonic oscillator is one of the great workhorses of physics. This is because physicists know how to solve this problem exactly. So, if they can get away with it, they always try to make the problem they are dealing with look like the harmonic oscillator problem. Can't blame them.

The simple pendulum is of course the most familiar example of the harmonic oscillator. This and a few other examples are illustrated in figure (a). If the system is initially at rest, then oscillations can be triggered off by displacing the particle from its equilibrium position. The oscillations can be described by the equation

$$x(t) = A \sin \omega t$$

where $x(t)$ is the amplitude at the time $t$, $A$ is the maximum amplitude and $\omega$ is the angular frequency ( $= 2\pi \nu$ ). The period $T$ of oscillation is given by $T = 2\pi/\omega = 1/\nu$. In real life, harmonic oscillator behaviour is never observed since all oscillators invariably have some friction which causes the oscillations to damp. Even if friction is negligible, the harmonic oscillator behaviour is not observed if the amplitude of oscillation is large.

The harmonic oscillator of the physicist ignores such complications—naturally. It is an idealised model, in which the potential energy varies as

$$V = (1/2)kx^2$$

where $k$ is called the spring constant and $x$ is the displacement. If the potential energy has this behaviour, then the equation of motion is

$$m\ddot{x}(t) = -kx(t)$$

where $m$ is the mass of the particle. The left-hand side of the above equation describes the force as given by Newton's law while the right-hand side describes the restoring force due to the spring action. It is easy to verify that the expression given earlier for $x(t)$ satisfies the above equation of motion if $\omega^2$ is chosen equal to $(k/m)$.

Oscillators are encountered in the world of atoms and molecules also. A classic example is provided by the hydrogen molecule which is formed by two hydrogen atoms coming together. They are held by an 'electronic glue' but for practical purposes, the two protons act as if they are held together by a spring. In turn this means that the atoms can jiggle, the jiggling being harmonic.

In the case of mechanical oscillators like the pendulum, we can sort of dictate its amplitude and also its energy (that is if we don't look too closely). Not so in the case of the oscillators of the atomic world. When there is a collection of such oscillators at a temperature $T$, they all organise their motions suitably to be indicative of certain averages corresponding to that temperature. What it means is the following:

1. All oscillators do not have the same amplitude; some have more and others have less.
2. We cannot say what the amplitude of a given oscillator is. We can at best talk about the *average* amplitude of an oscillator.
3. The average amplitude increases with temperature.
4. The average energy of a *classical* one-dimensional oscillator at temperature $T$ is $k_B T$; for a three-dimensional oscillator it is $3 k_B T$.

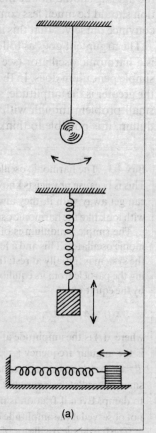

(a)

course would depend on its amplitude). In a solid we have $\sim 10^{23}$ tiny oscillators, and we can't say what their individual energies are. We can only say that on the average, each oscillator has an energy of so much; but knowing the average energy is enough.

Let us say the average energy per atom is $\varepsilon$. We further suppose that $\varepsilon < B$, the binding energy, since we want our solid to stay a solid. Can we say something more about this average energy $\varepsilon$? Sure we can, because that is related to the temperature $T$ of the solid—the higher the temperature, the greater is the value of $\varepsilon$. In fact, for a one-dimensional oscillator,

$$\varepsilon = k_B T \tag{4.3}$$

For a three-dimensional oscillator, we have similarly,

$$\varepsilon = 3k_B T \tag{4.4}$$

Let $b$ denote the amplitude (at temperature $T$) of our one-dimensional thermal oscillator or atomic oscillator, whatever you prefer to call it. If $\varepsilon = B$, the atoms would separate; this means that the amplitude of vibration will then be roughly half the interatomic spacing, i.e. $(a/2)$. So,

$$(a/2) \sim (B)^{1/2} \tag{4.5}$$

(Caution: If you are fussy about dimensions, then you must be careful for the dimension of $a$ is not the same as that of the square root of $B$. All that (4.5) says is that $a$ varies roughly as the square root of $B$.) In general the amplitude at $T$ is given by

$$b \sim (k_B T)^{1/2} \tag{4.6}$$

Taking the ratio,

$$b \sim (a/2) \cdot (k_B T/B)^{1/2} \tag{4.7}$$

Let us now put in some numbers. We take $B = 5$ eV and assume $T$ is the room temperature, i.e. $T = 300\ °K$. (300 is a nice and convenient number. So physicists always assume the room temperature to be 27 °C whether it is the North Pole or the Rajasthan desert. 27 °C means the absolute temperature is 300 °K.) Then $k_B T = (1/40)$eV, and the amplitude is

$$b \sim (a/30) \ll a \tag{4.8}$$

(the symbol $\ll$ means *much less than*). So we see that the vibration amplitudes at room temperature are indeed small compared to the lattice distance.

## 4.3 Evaporation temperature of solids

Let me begin this section by simply quoting Weisskopf, as it would also convey the flavour of his style. He says:

> One can take a handbook of constants, look up a lot of numbers such as binding energies, elasticity coefficients, melting temperatures and so on, and then ask oneself: 'Why are they that big and not a 100 times smaller or bigger?' This is the question we are now going to ask ourselves regarding the evaporation temperature of solids. These are usually of the order of 3,000 degrees. Let us now calculate them.

We now pursue this same question but in our own way, i.e. slow and easy. When we heat a solid, it first melts and then the liquid starts evaporating. Why? Because evaporation is a process where atoms take leave and fly away from the liquid. In other words, heat makes the atoms unbind.

Normally, we have the sequence:

solid → liquid → vapour.

However direct evaporation from solids is also possible. Called *sublimation*, you must have seen it happen with camphor and naphthalene balls. This is the sort of situation we consider now (Fig. 4.5a).

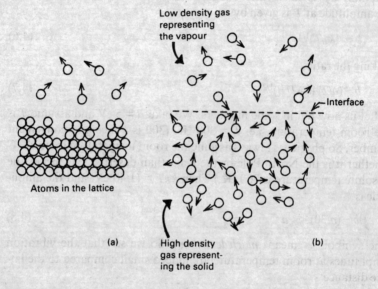

Fig. 4.5 Figures to illustrate Weisskopf's calculation of the evaporation temperature of solids. (a) illustrates schematically the evaporation process itself. In (b) is shown a convenient model. There are two gases of vastly differing densities in contact with each other across an interface. At the evaporation temperature, the numbers of atoms crossing the interface in the two directions are equal to each other.

You might say:'Haven't we said in the previous section that a solid will come apart if we give to each atom a kinetic energy such that

$$K \sim (1/2)k_B T \sim B ?'$$

Sure the solid would come apart. But if we use the above equation to estimate the evaporation temperature, then we would get evaporation/melting temperatures to be of the order of 50,000 degrees whereas we know they are much less. So the above method gives an *overestimate*.

Clearly we must tackle the problem differently to get more realistic numbers. The evaporation temperature depends on the pressure of the gas surrounding the solid. Let us assume that this pressure is one atmosphere.

We now need a model. A convenient one is illustrated in Fig. 4.5(b) in which we regard the solid as a very high density gas, i.e. a gas with a density 1000 times larger than that of the vapour above it. Next, we make the assumption that the number of molecules crossing the interface and going up in one second is the same as the number coming down and hitting the interface, i.e. the number escaping from the solid into the vapour per unit time is the same as that returning from the vapour into the solid (again per unit time). So what we have to do is to calculate these two rates and put them equal. How does this help? Well, it helps in the following way. You see the two rates become equal only at the evaporation temperature, and the value of the temperature enters in the rate for escape from the solid.

Suppose we have an atom at the top surface of the solid. The probability for it to tear off and escape into the vapour is (see Box 4.4)

---

**Box 4.4** In thermal physics, one often comes across a factor like

$$e^{-E/k_B T}$$

Usually, this refers to the probability for a particle to escape over a barrier of height $E$, as illustrated in the figure. Here, there is a particle in a local valley at the point A. If it is given a push, the particle will rattle around a bit about A, like an oscillator. Let us suppose that somehow the particle gets to the spot B. It then has a chance to roll down to the other side and escape. So the escape probability is essentially the probability to reach the spot B from the spot A. And this is what the expression above gives, i.e. it is the probability of escaping over a barrier of height $E$. What is the guiding hand that helps the escape? Thermal forces of course. It is presumed that the 'system' escaping is in contact with a heat bath (as it is called) which is at a temperature $T$; it is the heat bath or reservoir which supplies energy to the escaping fellow. Obviously, the higher the temperature, the easier is the escape.

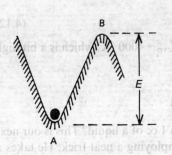

$\sim (e^{-B/k_BT})$

If $N_U$ atoms reach the surface per second from below, then the number of atoms escaping from the solid into the vapour above is equal to the number reaching the top, multiplied by the probability for one atom to escape from the top, i.e.

$$N_U (e^{-B/k_BT})$$

Now let $N_D$ be the number of atoms hitting the interface per second from above. We suppose that when the atom hits from the top, it gets into the solid. Then, according to our assumption above we can write

$$N_D = N_U (e^{-B/k_BT}) \tag{4.9}$$

Now what about $N_D$ and $N_U$? Here Weisskopf makes use of the model of Fig. 4.5(b) and says that $N_U$ and $N_D$ are very similar quantities, their values being different because the densities of the two gases are different. So he assumes,

$$N_U = 1000 \, N_D \tag{4.10}$$

With this we can now write

$$e^{-B/k_B T_{evp}} \sim 1/1000 \tag{4.11}$$

or

$$B/k_B T_{evp} \sim \log_e 1000 = 7$$

or

$$k_B T_{evp} = B/7 \tag{4.12}$$

Taking $B$ to be around 3 eV, we get $T_{evp} \sim 4000\,°K$, which is a bit high but of the right order of magnitude.

### 4.4 Number of atoms in 1 cc of a liquid

Roughly how many atoms are there in 1 cc of a liquid? This is our next target. Weisskopf estimates this by employing a neat trick. He takes a cube of a liquid, 1 cm × 1 cm × 1 cm. In a liquid, the atoms are not arranged in a regular fashion; rather, they are disordered. However, Weisskopf pretends that they are neatly arranged as in Fig. 4.6, and imagines the array to be systematically sliced as in Fig. 4.7. Thus in the end all the bonds are broken and we have separated atoms. This is the equivalent of completely evaporating 1 cc of liquid, and would require an energy equal to the latent heat of vaporisation which we call $E_{evp}$.

*On solids and liquids* 59

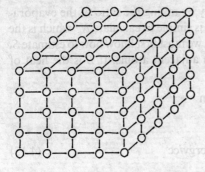

**Fig. 4.6** Weisskopf's model for calculating the number of atoms per cc in a liquid. The atoms are supposed to be arranged neatly on a lattice as shown, there being $N^3$ atoms per unit volume.

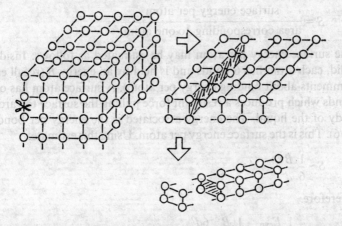

**Fig. 4.7** Disassembly of atoms in a volume 1cm × 1cm × 1cm of a liquid. The atoms are first imagined to be arranged in a lattice as in Fig. 4.6. The lattice is then systematically sliced as shown. Slicing exposes fresh surfaces. If $N$ is the number of atoms along each edge, then the total surface area exposed is $6N$.

Let us now examine the slicing business a bit more closely. You notice that the slicing process exposes fresh surfaces. This costs energy, i.e. to expose an area $A$, we must supply an energy of $A \times S$ where $S$ is the *surface energy*, being the energy needed to create 1 cm² of fresh area. It may take you a few minutes, but I am sure you can figure out that the total area created when all the bonds are snapped is ~ $6N$. This requires an energy of $6NS$. But this is also the heat of evaporation. Therefore,

$$E_{\text{evp}} = 6NS \qquad (4.13)$$

or

$$N = E_{\text{evp}}/6S \qquad (4.14)$$

Once we have calculated $N$ this way by taking the ratio of the evaporation energy to the surface energy, it is simple to calculate $N^3$ which is the number of atoms in 1 cc. But for this, we must first know how to estimate $S$.

Let $a$ be the separation between adjacent atoms. If $B$ is the heat of evaporation per atom, then

$$\frac{\text{heat of evaporation per atom}}{\text{volume per atom}}$$

$$= \frac{B}{d^3} = E_{\text{evp}} = \text{evaporation energy/cc} \qquad (4.15)$$

Turning to $S$, we can express it as

$$S = \frac{\text{surface energy per atom}}{\text{area corresponding to one atom}}$$

The surface energy per atom may be estimated as follows: Inside the solid, each atom has 6 bonds and is held by an energy $B$ (recall earlier comments about binding energy per atom). A surface atom has only 5 bonds which produces a resultant force pulling the surface towards the body of the liquid. The energy associated is the energy per bond, i.e. ($B/6$). This is the surface energy per atom. Using this estimate,

$$S \sim \frac{1}{6} \frac{B}{d^2}$$

Therefore

$$N = \frac{1}{6} \frac{E_{\text{evp}}}{S} = \frac{1}{6} \frac{B}{d^3} \cdot \frac{6d^2}{B} \sim (1/d) \qquad (4.16)$$

Since $d \sim 10^{-8}$ cm, we get $N \sim 10^8$ and $N^3 \sim 10^{24}$, which is OK.

Weisskopf adds:

> This is a wonderful way of counting atoms—wonderful because the number of atoms is terrific and yet it is reduced to two measurable human experiences. We know how much energy it takes to boil away 1 cc of water—we do it everyday. We also have a feeling for how much energy it needs to extend a surface, for example in blowing a soap bubble. Yet, together they give us a number so great that we cannot really visualise it.
>
> This should be in every elementary physics textbook, but it can only be found in one little-known book which is otherwise very bad! Written by a (fortunately) completely unknown German physical chemist and entitled 'German Chemistry', it came out at the height of the Nazi regime in Germany in 1938. It contained a lot of chapters that were strongly against quantum mechanics

(this being a Western, Jewish invention) and insisted that chemistry should be much simpler—and that German chemistry was, quoting this as an example. It shows that one can find pearls wherever one looks for them.

Perhaps, I should add a few words of comment. In the early thirties, the Nazis led by Adolf Hitler established a dictatorship. Hitler hated the Jews and began to persecute them. Many people therefore fled from Germany, and this included dozens of eminent scientists (one of whom of course was Weisskopf himself; Einstein was another—the list is long). Hitler declared that all scientific discoveries made by Jews were wrong, and this included the famous theory of relativity! The reputed German scientists who were not Jews were quite unhappy about this, but a number of third-rate scientists who wanted to please Hitler wrote a lot of nonsense. This is no doubt silly but not at all bad compared to what Hitler himself did during the height of the Second World War—he arranged for the systematical killing of *six million* Jews left behind in Germany and Poland. Jews feel bitterly about this and are unable to forget it, even fifty years later. But notice how Weisskopf, though of the Jewish faith himself, is generous enough to talk about finding pearls. That speaks for the man.

# 5  Height Of Mountains

We all know that Mount Everest is about 10 km high. It is also the highest mountain in the world. Have you ever wondered why there are no mountains higher than Everest? You may say: 'Look, the Himalayas were formed due to a big collision between two huge landmasses (did you know that?). Mountain formation depends on geological happenings. We don't have higher mountains because a collision of such magnitude did not ever take place.' Well, you would be right of course but I could still ask: 'Suppose I can *build* a mountain; how tall can I make it? Is there any limit to its height?'

This is an interesting question and is the one Weisskopf (implicitly) poses and answers. What he does is really cute for we can use the same arguments to estimate the height of mountains on Mars, Jupiter, neutron stars and what have you. Incidentally, there is a somewhat related type of question: How tall can a tree grow? There are some differences between this question and the one we are concerned with at present. Anyway, you can think about it later.

OK. So our question is: 'Why don't mountains grow infintely high?' Simple; if a mountain is *very* high, it would start sinking. What really happens during the sinking process is that the weight of the mountain causes the atomic and molecular bonds in the rock below to break. This makes the underlying rock melt and flow aside so that the mountain as a whole starts sinking—see Fig. 5.1.

We can now do some estimating. We first say the mountain sinks by a certain amount; this results in a certain loss of gravitational energy like when a stone falls. Where does this energy disappear? It is used for liquefying the rock below. So you know what we have to do. We have to calculate these two quantities namely, the loss in gravitational energy and the energy required for melting, and then equate the two.

Let the mass of the mountain be $M$, its height $h$, and let it sink by a distance $x$—see Fig 5.1. This figure may shock you, but it is quite in the spirit of the back-of-the-envelope game. The loss in gravitational energy is given by

$$Mgx \tag{5.1}$$

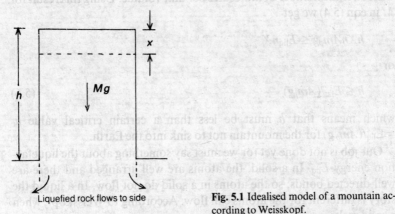

**Fig. 5.1** Idealised model of a mountain according to Weisskopf.

If the mountain has to sink by a distance $x$, then obviously, a layer of thickness $x$ below the mountain must melt and flow away. If $X$ is the area of cross-section of the base of the mountain, then the energy needed to liquefy this layer is given by

$$E_{\text{liq}}(n \cdot x \cdot X) \tag{5.2}$$

where $n$ is the number of molecules per unit volume and $E_{\text{liq}}$ is the liquefaction energy or the latent heat of melting per molecule. So we equate the above two quantities to obtain

$$Mgx = E_{\text{liq}}(nxX)$$

Cancelling $x$ then gives

$$Mg = E_{\text{liq}} nX \tag{5.3}$$

What eqn (5.3) says is that in order for the mountain to sink, its mass $M$ must have a certain minimum value. If $M$ is less than this value, then the mountain will not sink. Therefore the masses of stable mountains are given by

$$Mg \leq E_{\text{liq}} nX \tag{5.4}$$

Let $m$ denote the mass of a molecule of rock, and $A$ be the number of protons and neutrons in the molecule. If $m_p$ denotes the mass of a proton, then $m = Am_p$ (remember, mass of the neutron is *approximately equal to* mass of the proton). Thus we can write

$$M = \text{vol. of mountain} \times (\text{no. of molecules/unit vol.}) \times \text{mass of molecule}$$
$$= hX\, nAm_p \tag{5.5}$$

where $n$ is the number of molecules per unit volume. Using this result for $M$ in eqn (5.4) we get

$$hXnAm_pg \leq E_{liq}nX$$

or

$$h \leq E_{liq}/(Am_pg) \tag{5.6}$$

which means that $h$ must be less than a certain critical value $h_c = E_{liq}/(Am_p g)$ for the mountain not to sink into the Earth.

Our job is not done yet for we must say something about the liquefaction energy $E_{liq}$. In a solid, the atoms are well arranged and there are well-directed bonds; so the atoms in a solid do not flow. In a liquid the story is different for the atoms can flow. According to Weisskopf, when a solid melts,

> the whole bonds between the atoms are not broken, just the directionality of the bonds. This enables a liquid to flow, whereas a solid cannot because its bonds are held in fixed positions relative to the constituent atoms. The energy necessary to break the directionality of a bond, i.e. to liquefy, is evidently less than the binding energy. It is difficult to estimate just how much less, because the theory of the liquid state is not very well developed.

OK. So we have to do a bit of guesstimating. Weisskopf does this by comparing the latent heat of melting for ice (which is ~ 80 cal/gm) with the latent heat of vaporisation (which is ~ 560 cal/gm). The ratio of the latent heat of melting to that of vaporisation is (1/7) in this case. We assume this ratio applies to all materials, not merely to water and ice. We have to somehow link these things up with the binding energy.

Now the total binding energy is the energy we must supply to a solid to tear it completely apart into separated atoms. Surely this must be somewhat more than the energy required to tear a *liquid* apart into separated atoms. (Remember that the energy required to tear a liquid apart is nothing but the latent heat of vaporisation). So Weisskopf takes

energy of melting ~ (1/7) energy of vaporisation

~ (1/10) total binding energy.

So we may write

$$E_{liq} = \beta B \tag{5.7}$$

where $\beta \sim 0.1$ for reasons just given (remember that both $E_{liq}$ and $B$ are defined per molecule). As for $B$, its value for silicon dioxide (the main constituent of rock) is about 2.7 eV. So,

$$h \leq (0.1 \times B)/(Am_p g) \tag{5.8}$$

Substituting numbers, one gets

$$h \leq 30 \text{ km} \tag{5.9}$$

Weisskopf says:

> This shows that a mountain must be less than 30 km high to be supported by the rock at its base. Actually, the upper limit is smaller than that because the rock is warm and therefore needs less energy to liquefy. All the mountains we find on earth are of a height which is of this order or smaller. On other planets the critical height would be different, and the planet may also be made of different material.

I am told that on neutron stars, the mountains would be only a few centimetres high.

Some cleaning up is required since we want an expression for $h$ in terms of fundamental constants alone. Let us get rid of $g$ because it is not a fundamental constant. We know how to do this because $mg$, the force of attraction between a mass $m$ and the Earth, is given by

$$mg = GmM_E/R_E^2 \tag{5.10}$$

where $M_E$ is the mass of the Earth and $R_E$ is its radius. Cancelling $m$ on both sides,

$$g = GM_E/R_E^2 \tag{5.11}$$

Now $M_E$ and $R_E$ can be expressed in terms of $N_E$, the number of nucleons in the Earth. The Earth consists mostly of silicon dioxide (for which we suppose $A = 60$) and iron which forms the core ($A = 67$). The silicon dioxide molecule and the iron atom have approximately the same value for $A$. So, the number of molecules equals $(N_E/A)$. Also, we can take that iron and silicon dioxide both have the same radius which we write as $\sim fa_0$ with $f \sim 4$.

Now the radius of the Earth can be expressed in terms of the radius of the individual molecules and the number of molecules in the Earth. Since the volume of Earth $= (4/3)\pi R_E^3$, crudely,

$$\text{vol.} \sim R_E^3 = (\text{no. of molecules}) \cdot (\text{vol. of molecule})$$
$$= (N_E/A) \times (fa_0)^3 \tag{5.12}$$

From this,

$$R_E = (N_E/A)^{1/3} (fa_0) \qquad (5.13)$$

Use this in eqn (5.11) and simplify; we then get

$$g = GN_E m_p (A/N_E)^{2/3} (1/fa_0)^2 \qquad (5.14)$$

which may be substituted in eqn (5.8) to obtain an expression for the critical height $h_c$ in terms of the fundamental constants.

We could really stop here but Weisskopf would like to take things a bit further. He wants a formula which contains only dimensionless constants. $G$, the gravitational constant has dimensions. This he gets rid of as follows. The gravitational interaction energy between two nucleons is given by

$$Gm_p^2/r \qquad (5.15)$$

where $r$ is the distance between them. This of course is similar to the formula for electrostatic interaction between two particles of charge $e$ for which the expression is $e^2/r$. From this we see that $e^2$ and $Gm_p^2$ play similar roles. This may be exploited to define a fine structure constant for gravity similar to the usual fine structure constant, i.e. we write

$$\alpha_G = Gm_p^2/\hbar c \qquad (5.16)$$

This quantity is dimensionless, and has the numerical value

$$\sim 0.6 \times 10^{-38} \qquad (5.17)$$

Compare this with $\alpha$ which has a value of $\sim (1/137)$. This indicates how weak gravity is. See also Box 5.1.

If $G$ is written in terms of $\alpha_G$, then we obtain the following expression for $g$ in terms of fundamental constants:

$$g = (\alpha_G \hbar c/m_p) \cdot A^{2/3} \cdot N_E^{1/3} \cdot (1/fa_0)^2 \qquad (5.18)$$

---

**Box 5.1** Weisskopf gives the following example to illustrate how weak gravity is compared to electromagnetic forces. He borrows the example from the Israeli scientist Y. Ne'eman.

Consider the Apollo rocket. This was the one used for putting man on the Moon. It is rather similar to the rocket used nowadays for sending up the space shuttle; you might have seen launches of the space shuttle on TV. Consider 1 mm³ of iron in the centre of the rocket. Imagine that all the electrons in this tiny piece of iron are removed and placed below the rocket. The rocket would now be positively charged, and roughly fifty metres below it would be present an equal amount of negative charge. Naturally the two charges would

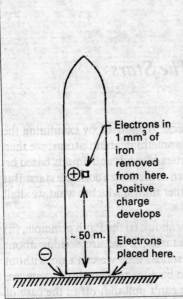

attract each other. This attraction is so strong that it can prevent the rocket from lifting off! In other words, the force of electrostatic attraction is $\sim Mg$, where $M$ is the mass of the whole rocket.

Feynman gives a similar illustration. He considers two humans standing about one metre apart. Somehow, each of them has 1% more electrons in their bodies than protons. This of course does not happen in normal human bodies, but let us say that these two fellows manage to acquire excess negative charge. Naturally there would be a repulsive force between the two. Feynman asks: 'How great? Enough to lift the Empire State Building? No! To lift Mount Everest? No! The repulsion would be enough to lift a weight equal to that of the entire Earth!'

Substituting for $g$ in (5.8) and remembering that the fine structure constant $\alpha = (e^2/\hbar c)$ then gives for the critical height $h_c$ the result

$$h_c = (0.1 \times B) f^2 \cdot (1/\alpha_G) \cdot (1/N_E)^{1/3} \cdot (1/A)^{5/3} \cdot a_0^2 \qquad (5.19)$$

Let $B = \eta$ Rydbergs. Remembering $a_0 = (\hbar^2/m_e e^2)$ and that 1 $Ry$ = $(m_e e^4/2\hbar^2)$, we finally obtain

$$\frac{h_c}{a_0} \sim (0.1 \times \eta) f^2 (\alpha/\alpha_G) (1/N_E)^{1/3} (1/A)^{5/3} \qquad (5.20)$$

This is the desired expression for $h_c/a_0$ in terms of dimensionless constants. Observe that only $N_E$ varies from planet to planet. So, once we know the number of protons in any planet, we can calculate $h_c$ for that planet. Of course, we can't use this formula for a celestial object with a different material structure since $E_{\text{liq}}$ in that case would be different. In this case, we have to go back to eqn (5.6). If one does this for a neutron star, one gets an unbelievable result. Try to find out how high the tallest mountains there are.

# 6   On The Way To The Stars

Let us pause for a minute and look back. We started by examining the energy levels of atoms. Having learnt something about atoms, we then tried to understand some of the properties of solids and liquids based on atomic properties. We now wish to learn something about the stars. But this requires that first we learn a few other things, which is what we shall be doing now.

There are three things we must know about: (i) the Pauli principle, (ii) radiation and (iii) nuclear forces. Actually, if we are serious about understanding atoms, molecules and solids, then we can't do without the Pauli principle. But for playing the back-of-the-envelope game, we could manage without it; however, we can't pull that off in the case of the stars. The same with radiation. True we generate radiation when we heat something or do a radio or TV transmission. Also, without radiation from the Sun, we would be nowhere. Still, we can understand many things happening on the Earth without bringing in radiation, for example the height to which the mountains can rise. But you just can't discuss a star without bringing in radiation. Once again, it is the same with nuclear energy. If you forget radioactivity, nuclear reactors and the hydrogen bomb, then you really don't need nuclear physics to understand most of what is going on in our neighbourhood. Not true in the case of the stars since energy production there depends very much on nuclear forces. I always find this very remarkable—I mean the stars are such huge objects and yet what goes on inside them is directly dependent on some of the tiniest things we can think of in this Universe.

## 6.1   The Pauli principle

The Pauli principle is one of the cornerstones of physics. It applies to some objects like the electron, proton and the neutron, but not to some others like the photon and the helium atom. When does it apply and when does it not? The short answer is that the principle applies if the particles are of the Fermion type (Box 6.1). There is yet another species of particles called Bosons, and the principle does not apply to them.

**Box 6.1** Elementary particles have many attributes like mass, charge, etc. One quantity in this list is the so-called *spin*, a better name for which is *intrinsic angular momentum*. Soon after Bohr came up with his atom model, it was believed that the electron while going round the nucleus in a circular orbit was also spinning about an axis of its own, rather like our Earth does while going round the Sun. Later, after the birth of quantum mechanics, it was realised that this picture of electrons orbiting had no validity; likewise also, the picture of electrons spinning about an axis. However, the electron *did* possess an intrinsic property which people earlier had tried to picture in terms of an electron actually spinning about an axis of its own. For the sake of historical continuity, this property was called *spin*, though it did not mean the electron was spinning. This property enabled the electron to be in two different states. Pictorially, one represents these two states by drawing an arrow pointing either up or down. Correspondingly, one talks of a 'spin-up' electron or a 'spin-down' electron. If a magnetic field is present, these two spin states have different energies.

Soon after the discovery of electron spin, it was realised that many other particles like the proton, the neutron, etc., also possessed this property of spin. It was further found that spin could be characterised in terms of a number that was either a half-odd integer (i.e. 1/2, 3/2, 5/2, ...) or an integer (i.e. 0, 1, 2, 3, ...). In the zoo of elementary particles, these two types form two species. The former are called *Fermions* (after Enrico Fermi), while the latter are called *Bosons* (after Satyendranath Bose).

In effect the Pauli principle says that Fermions are very shy and generally try to avoid each other. There are many ways of stating the principle; one of these is:

*It is impossible to have two or more identical Fermions in the same quantum state.*

The meaning of this is explained in Box 6.2. There is a fancier statement involving wave functions and all that. This wave function business is very important and at the heart of quantum mechanics. Maybe you don't know about them but don't worry. Box 6.3 will tell you what you need for the present.

**Box 6.2** Wolfgang Pauli pointed out that a collection of identical Fermions (e.g. of electrons, or of protons, etc.,) have an interesting property namely, that in a given energy state, there can be only one Fermion. In the case of the electrons, if the spin-up and the spin-down energy levels coincide (as frequently happens), then there can be two electrons in that level, one with spin-up and another with spin-down. Note that one *cannot* have both electrons with their spins up or, for that matter, both with their spins down—see figure. In short, there is very limited occupancy in each energy level; others are kept out, and this is the essence of the *exclusion principle*. A more general statement of the above is discussed in Box 6.4.

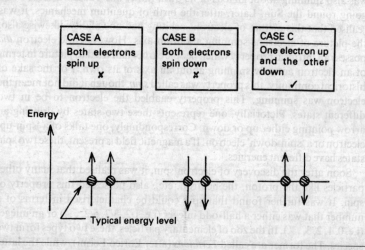

**Box 6.3** In quantum mechanics one deals with wave functions, usually denoted by the symbol $\psi$. The nature of the system determines the variables on which $\psi$ depends. If there is only one particle, then one writes the wave amplitude as $\psi(x)$ to denote the wave amplitude at the point $x$. Two things about the wave function; (i) it is a complex quantity, and (ii) it cannot be measured by any experiment. However, the quantity

$$|\psi(x)|^2 \, dx = \psi(x)\psi^*(x)dx$$

(where $\psi^*$ denotes the complex conjugate of $\psi$) does have a clear physical meaning and is measurable. It denotes the probability of finding the particle between $x$ and $x + dx$.

How does one determine the wave function of a given system? For that one must solve an equation called the *Schroedinger equation*—not always easy! But for a particle not subject to any force, i.e. a free particle, the solution is easy and is given by

$$\psi(x) = \text{constant} \times e^{ikx} \tag{a}$$

The quantity $k$ is called the wave vector, and its dimension is $L^{-1}$.

In the above, spin has not been introduced but it has to be if one is to be realistic. How that is done (formally) is indicated in Box 6.4. In the text we work with a wave function like (a) and reason out the effects due to spin.

You might be wondering why on earth we are bothering about this Pauli principle and what it has got to do with stars. Perhaps I should give you a hint. You see, stars contain lots and lots of electrons which, as I just told you, obey the Pauli principle. So what? Well, when electrons try to avoid each other, it is as if an outward pressure is acting on the collection of electrons (sometimes called the electron gas). Such a pressure is also called the *degeneracy pressure*. In short, the Pauli principle is responsible for degeneracy pressure inside stars.

Earlier I hinted that the Pauli principle can be stated in other ways. One involving wave functions is:

*The wave function of a pair of identical Fermions (e.g. both electrons) must be antisymmetric under the exchange of the two Fermions.*

The meaning of this is explained in Box 6.4.

Let us now take two electrons of the same energy. We assume for simplicity that the motion of the electrons is restricted to one dimension. Let their positions and momenta be $x_1, \hbar k_1$ and $x_2, \hbar k_2$, respectively. The quantity $k$ multiplying $\hbar$ is called the *wave vector*. Typically, $k \sim 2\pi/\lambda$, where $\lambda$ is the de Broglie wavelength.

The wave function of the pair must be the product of the wave functions of the individual electrons. For a free electron the wave function is a plane wave and is written $e^{ikx}$. So for a pair we have

---

**Box 6.4** Let $\psi(x)$ denote the wave function of a particle without reference to its spin. The variable $x$ denotes the space coordinate. With spin included, let us denote the wave function as $\psi(x, \sigma)$ where $\sigma$ is the spin coordinate. In the case of a spin 1/2 particle, this coordinate can take two discrete values namely $+(1/2)$ and $-(1/2)$; similar values can be assigned in the case of other spins. While $\sigma$ can take only a limited set of values, the coordinate $x$ can vary continuously over the whole of space.

What we have so far considered is the one-particle wave function. When two particles are present, the wave function is written $\psi(x_1, \sigma_1; x_2, \sigma_2)$. Similarly, the three-particle wave function is written $\psi(x_1, \sigma_1; x_2, \sigma_2; x_3, \sigma_3)$, and so on. For compactness let us write 1 for $x_1, \sigma_1$, 2 for $x_2, \sigma_2$, etc. Then the wave functions for varying number of particles could be represented as $\psi(1)$, $\psi(1,2)$, $\psi(1,2,3)$, etc.

Consider now a two-particle system represented by the wave function $\psi(1,2)$. Suppose that we interchange or *exchange* all the coordinates of the two particles. This results in a new wave function $\psi(2,1)$. Question: How is $\psi(1,2)$ related to $\psi(2,1)$? It turns out that if the two particles are Bosons, then $\psi(1,2) = \psi(2,1)$ (whereas if the two are Fermions, then $\psi(1,2) = -\psi(2,1)$. In the former case the wave function is said to be symmetric and in the latter case it is said to be antisymmetric. Similarly, in a three-particle system,

$$\psi(1,2,3) = +\psi(3,2,1) \text{ Bosons},$$

while

$$\psi(1,2,3) = -\psi(1,3,2) = +\psi(3,1,2)$$
$$= -\psi(3,2,1) \text{ Fermions}$$

The idea of symmetry and antisymmetry of wave functions can be extended to $N$-particle wave functions.

The *generalised* Pauli principle says:

Take the $N$-particle wave function $\psi(1,2,...i,...j,...N)$. Exchange particles $i$ and $j$. Call the new wave function $\psi'$. If the $N$ particles are Bosons, then $\psi' = \psi$. If they are Fermions then $\psi' = -\psi$.

Why this behaviour? Ah, that is a deep one. It is all connected with relativity, and was first explained by Pauli. About this connection, Richard Feynman says: 'It appears to be one of the few places in physics where there is a rule which can be stated very simply, but for which no one has found a simple and easy explanation. The explanation is deep down in relativistic quantum mechanics. This probably means that we do not have a complete understanding of the fundamental principle involved. For the moment, you will just have to take it as one of the rules of the world.'

$$\psi = e^{ik_1 x_1} e^{ik_2 x_2} \qquad (6.1a)$$

By going over to the centre-of-mass description (see Box 6.5), one can write the wave function as

$$\psi = e^{ik(x_1 - x_2)}$$
$$= e^{ikx} \qquad (6.1b)$$

where $\hbar k_1 = -\hbar k_2 = \hbar k$ is the momentum of each electron in the centre-of-mass frame, and $x = x_1 - x_2$ is the separation of the two electrons. This expression for the wave function is, however, not completely OK because if we swap the positions of the two electrons, i.e. if we make $x_1$ into $x_2$ and vice versa, then, according to the antisymmetry principle, the wave function must change sign. But what we have above does not. The correct wave function turns out to be

$$\psi = e^{ikx} - e^{-ikx} \qquad (6.2)$$

It is easy to see that if the positions of the two electrons are interchanged, then $\psi$ becomes $-\psi$.

On the way to the stars 73

**Box 6.5** Consider a collision between two balls, one of mass $m_1$ and which is stationary, and another of mass $m_2$ which is moving. Figure (a) shows a motion picture of the collision as we would see it. Let us call this the 'laboratory picture'. The crosses in the figure show the positions of the

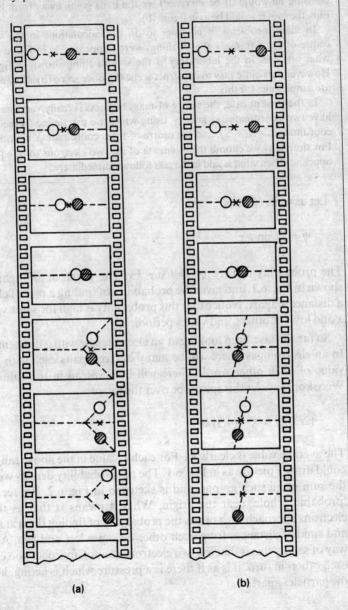

(a)          (b)

> centre-of-mass or, if you wish, the centre-of-gravity (the former is better). Obviously, the centre-of-mass is moving all the time.
>
> Suppose there is an observer riding with the centre-of-mass. Before the collision, he would see the two balls rushing towards him, while after the collision he would (if he survives!) see the balls going away from him. For him, the movie would be as in figure (b).
>
> In many problems, it is easier to do the calculations in the so-called centre-of-mass frame, i.e. as if things were happening as in figure (b). But what one sees in the laboratory is the sort of thing shown in figure (a). However, it is quite easy to construct scene (a), once scene (b) is known; there are simple rules for this.
>
> In the present case, the centre-of-mass business is really very simple. We have two coordinates $x_1$ and $x_2$, using which we can define (i) the relative coordinate $x = x_1 - x_2$, and the centre-of-mass coordinate $X = (x_1 + x_2)/2$. For simplicity we choose the momenta of the two electrons to be equal and opposite. Then what is said in the text follows immediately.

Let us write (6.2) as

$$\psi = 2i \sin kx \tag{6.3}$$

The probability density $|\psi|^2 = 4 \sin^2 kx$. The graph of $|\psi|^2$ against $x$ is shown in Fig. 6.1, and gives the probability of finding a pair of electrons a distance $x$ apart. Notice that this probability is high for some values of $x$ and low for others, and varies periodically.

So far we have been looking at an electron *pair* with a momentum $\hbar k$. In an electron gas, there will be many electron pairs each with its own value of $k$; in other words, there will be a spread in the values of $k$. Weisskopf says let this spread be over the range

$$(2/3) k_0 < k < (4/3) k_0$$

The average value is clearly $k_0$. For each $k$ value in the above range, one could draw a picture as in Fig. 6.1. The net probability density would be the sum of all such graphs, and is sketched in Fig. 6.2. We see here a 'probability hole' near the origin. What it means is that as the two electrons approach each other, the probability of finding them at smaller and smaller distances from each other becomes less and less. A better way of saying this is that the two electrons cannot approach too close to each other; in turn, it is as if there is a pressure which is acting, keeping the particles apart.

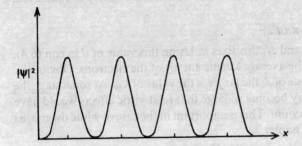

**Fig. 6.1** Probability density $|\psi|^2$ for an electron pair. The graph shows the probability of finding a pair of electrons having momenta $\hbar k$ and $-\hbar k$, and a distance $x$ apart. $\Psi$ is given by equation (6.3).

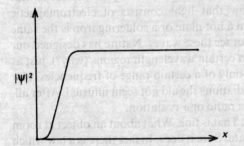

**Fig. 6.2** Cumulative probability of finding two electrons at a distance $x$ apart. This graph is obtained by superposing several curves of the type in Fig. 6.1, but corresponding to different values of $k$. Observe that the probability is low when $x$ is small.

Consider a gas made up of $N$ electrons. (An aside; a Pundit would not approve of this, for in such a gas, the electrons would immediately fly away on account of the electrostatic repulsion. There is nothing to keep them together. So while introducing an electron gas, one must always have a background compensating positive charge.) The electrons in such a gas will have many energy states. Correspondingly, they would have a spectrum of energies. Let $K_{av}$ denote the average kinetic energy of an electron in the gas, and $d$ denote the average spacing. Using $d$ we can define an average wave vector $k_{av}(= 2\pi/d)$ and also a related average momentum $p_{av}(= \hbar k_{av})$. Thus,

$$K_{av} = p_{av}^2/2m_e$$
$$\sim \hbar^2/2m_e d^2 \qquad (6.4)$$

The exact answer for the average kinetic energy is

$$K_{av} = 5.5 \times \hbar^2/2m_e d^2 \qquad (6.5)$$

which comes out from the theory of the electron gas. So we are not doing too badly! How does one estimate $d$? That is easy. First recall the meaning of $d$. Now associate with each electron, a sphere of radius $d/2$. Then,

$$V = N\,(4/3)\,\pi\,(d/2)^3$$

Since we know $V$ and $N$, this fixes $d$. Using this value of $d$ in eqn (6.4), one can estimate the average kinetic energy of the electrons. Obviously, the smaller the value of $d$, the larger is the value of $k_{av}$. At some stage the momentum $p_{av}$ may become so large that relativistic effects would have to be taken into account. This is important in the case of white dwarfs, as I shall narrate in Chapter 7.

## 6.2 Blackbody radiation

You must be aware that when an object is heated it radiates; the soldering iron is a familiar example. What the object radiates is heat energy. You probably know that light consists of electromagnetic waves. What comes out from a hot plate or a soldering iron is the same sort of thing—only we cannot see these waves. Nature has designed our eyes so that it can see only in certain wavelength regions (why?), just as our ear can hear vibrations only of a certain range of frequencies. Not seeing all electromagnetic radiations should not seem unusual. After all, we don't see the waves from a radio or a TV station.

OK, so hot bodies radiate. That is fine. What about an object at room temperature? Does that radiate? Sure it does. In fact there is a law which describes this emission process. It is called the Stefan–Boltzmann law. The essence of that law is that the radiation is emitted at a rate proportional to $T^4$ where $T$ is the absolute temperature of the radiating body.

Suppose you have a beaker of water in front of you. The water is at a temperature above 0 °K. The beaker should therefore radiate away all the heat energy contained in it and the water should freeze; but it doesn't. Why? Think about it.

Let us get back to the Stefan–Boltzmann law. See Fig. 6.3. This illustrates the experiment usually done (in colleges) to verify the above law. If there is an opening of area 1 cm² in the heated cavity, the amount of heat coming out of it per second is given by

$$I = \pi^2 (k_B T)^4 / (5 \hbar^3 c^2) \tag{6.6}$$

This is the Stefan–Boltzmann law.

Suppose the aperture has a shutter in front of it. Let us open it for just one second and then close it. Radiation from the cavity will stream out and fill a certain volume. Clearly this volume is $c$, and the energy filling it is $I$. So the energy density or the energy per unit volume would be $(I/c)$. This is the formula Weisskopf is after, by his own methods of course.

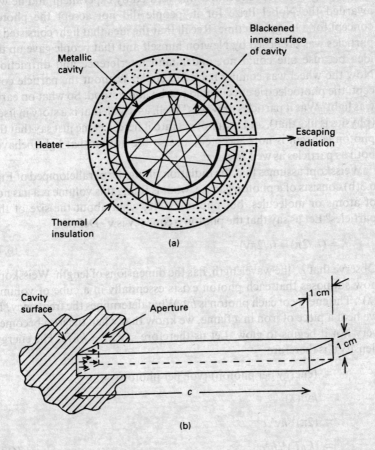

**Fig. 6.3** In the experiment done in college one has a hot spherical cavity and measurements are made on the radiation streaming out of it through a small aperture. (a) shows a schematic drawing of the blackbody. The hollow spherical cavity is blackened on the inside, suitably insulated and heated. The radiation trapped inside has a blackbody spectrum. A small portion of this radiation is allowed to leak out for observation—see (b). If the aperture is opened for one second, the radiation would fill a parallelopiped of length $c$.

The derivation he gives is really simple and he proceeds roughly as follows: Let us regard the radiation as composed of photons. I suppose you are aware that electromagnetic energy is packaged in tiny little energy packets called the *photon* (—Einstein who first gave us this idea actually called it the light quantum; the word 'photon' came much later). Historically, the idea of the radiation quantum was introduced by Einstein to explain the photoelectric effect, but the funny thing is that even though Einstein's formula was verified by experiment and he was awarded the Nobel Prize for it, people did not accept the photon concept for a long, long time. Recall that the idea that light consisted of particles was proposed by Newton himself and that people gave up the idea because one could not understand interference and diffraction. Now the wheel was coming back full circle. Without the particle concept, the photoelectric effect could not be understood. So what on earth was light? Was it particles or waves? Well, as usual, that is a story in itself (physics is like that), and I cannot get into it here. Let me just say that the photon concept is now accepted, as also the notion that light behaves both as particles as well as waves.

Weisskopf assumes that the radiation filling the parallelopiped of Fig. 6.3(b) consists of a photon gas, i.e. what is filling this volume is a gas not of atoms or molecules, but photons. OK, what about the size of the particles? Let us say that the photon frequency is $\nu$. Define

$$\bar{\lambda} = (\lambda/2\pi) = (c/2\pi\nu) \tag{6.7}$$

Observe that $\bar{\lambda}$, the wavelength, has the dimensions of length. Weisskopf now supposes that each photon exists essentially in a cube of volume $(\bar{\lambda})^3$. The energy of each photon is $h\nu$. What determines the frequency? If we heat a piece of iron in a flame, we know that when the iron becomes very hot it begins to glow. Let us therefore say $h\nu \sim k_B T$. The energy density $\varepsilon$ is therefore

$$\begin{aligned}\varepsilon &= \text{(energy per photon)/(vol. per photon)} \\ &\sim h\nu/(\bar{\lambda})^3 \\ &= (2\pi)^3 \, h\nu^4/c^3 \\ &= (k_B T)^4/\hbar^3 c \end{aligned} \tag{6.8}$$

which, apart from a few numerical factors, is the same as the energy density $(I/c)$ derivable from the Stefan–Boltzmann law. But we got this result in just a few steps.

## 6.3 Nuclear forces

As already mentioned, energy production in stars is governed by nuclear forces. Before we start talking about them, let us try to recall a few facts about atomic forces. First of all, what do I mean by atomic forces? An atom consists of a nucleus and some electrons, the whole thing being held together by electrostatic forces. Next, let us consider a molecule. A molecule is built-up by bringing together a few atoms. What sort of forces hold the atoms together? They are rather complicated, and we have already had a glance at them. One thing we can definitely say is that they are composite or synthetic forces, built-up essentially from the electrostatic forces between the various electrons in the different atoms. The complications arise because there are so many electrons and also because they receive orders from the various nuclei, as also from each other.

Let us now look inside the nucleus. The nucleus consists of a certain number of protons and a certain number of neutrons. In this environment, two nucleons (nucleon is the comprehensive name given to the proton and the neutron) interact with each other roughly as sketched in Fig. 6.4. Recall what I said a short while ago about interatomic forces, namely that it is a complicated left-over force from the coulomb interactions between the electrons of the constituent atoms. It could well be

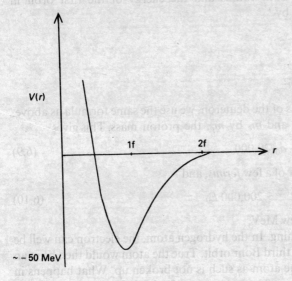

**Fig. 6.4** Schematic plot of the nucleon–nucleon potential. The shape of the curve looks roughly similar to the potential curve for two atoms like argon, only the distance and the energy scales are different.

that the nuclear force is also a derived or a left-over force. Derived or left-over from what? Perhaps the quarks. Have you heard of quarks? If not, try to find out something about them. You will find it very interesting. I am afraid I can't go into it here—maybe I will write about it separately. For the moment, let us just say that we *do* know that nuclear forces as shown in Fig. 6.4 exist, although we know little about their origin.

Let us use the form sketched in Fig. 6.4 to estimate the radius and the binding energy of the deuteron. Know what a deuteron is? It is a nucleus which consists of one proton and one neutron. The nuclear charge is 1. Therefore, the proton and the deuteron are isotopes (you must have studied about isotopes).

To do the estimation I just mentioned, we assume that the tail of the nuclear potential is given by a coulomb-like law, i.e. we suppose it varies like $\kappa^2/r$, where $\kappa$ is a quantity similar to the electronic charge, and is a measure of the strength of the nuclear force. At a distance of 1 fermi, the value of the electrostatic energy ($= e^2/r$) between two electrons is roughly 1 MeV. Weisskopf says that to get a rough comparison, one may assume $\kappa^2/r \sim 10$ MeV. This gives $\kappa \sim 3.3\, e$.

Recall that the Bohr radius and the energy of the first orbit in hydrogen are given by

$$a_0 = \hbar^2/m_e e^2$$

and

$$E_0 = -m_e e^4/2\hbar^2$$

To obtain the radius of the deuteron, we use the same formula as above, but replace $e$ by $\kappa$ and $m_e$ by $m_p$, the proton mass. This gives

$$a_D = \hbar^2/m_p \kappa^2 \sim a_0/20{,}000 \qquad (6.9)$$

which is of the order of a few *fermis*, and

$$E_D = -m_p \kappa^4/2\hbar^2 \sim 200{,}000\, E_0 \qquad (6.10)$$

which is around a few MeV.

One interesting thing. In the hydrogen atom, the electron can well be in the second or the third Bohr orbit. True the atom would then be in an excited state, but the atom as such is not broken up. What happens in the case of the deuteron? Let us suppose the deuteron existed in the first excited state. By analogy with the hydrogen atom, its radius would be about $4a_D$. But the nuclear force is practically zero at this distance. Consequently, the neutron and the proton would fly apart; in other words, the deuteron would become unbound! So, for the proton–neutron system, there is only *ONE* bound state possible, unlike in the hydrogen atom.

# 7 Something About Stars

We now leave Mother Earth behind, and move onto the world of stars. Presently, we shall make some estimates about the sizes of stars. The size of a star is determined by the forces acting in it and so let us look at them first.

## 7.1 The forces at work

A star's life is dominated by a perpetual struggle between two opposing forces, namely, gravity and pressure. Gravity tries to shrink the star while pressure tries to expand it. If gravity dominates, the star simply collapses while if pressure wins, the star goes on expanding; sometimes it explodes. If the two forces balance each other, the star is stable, at least for the while. To study the equilibrium of stars, we must obviously write down the expressions for the various forces at work and then do the balancing act.

Gravitational force and its consequences are easy to visualise. Every particle attracts every other particle and as a result they all like to come together as closely as possible. In fact, if there were no opposing forces, the star would collapse to a point! More about this later on. But here we note that this collapse does not always occur; what arrests it is pressure.

The fact that a star has pressure should not be a surprise since it is a mass of hot gas. But calculating this pressure can be tricky. You may ask: 'What is the problem? Can't we use the formula $PV = RT$ to estimate the pressure?' Not always; this formula is true only for an ideal classical gas. Stars contain a lot of electrons, and when electrons come close to each other, quantum effects become important. In addition, one has to watch out for relativistic effects. The story is not over yet. You see, there is a lot of light in a star. The atoms are busy emitting light, and if they did not, there would be no twinkle twinkle business. This radiation exerts pressure which also must be taken into account. In short, one has to be careful.

## 7.2 A bird's-eye view

Let us first have a bird's-eye view of the various estimates Weisskopf races through. I have already said that a star can be visualised as a mass of hot gas. In addition, Weisskopf supposes it is made up of $N$ protons and $N$ electrons (i.e. $N$ hydrogen atoms, completely ionized).

To start with we imagine the cloud to be big and diffuse. Gravity would make the cloud shrink, and in the process also heat it up. Initially one can use the perfect gas law $PV = RT$ to describe what is going on. But when, as a result of the shrinking, the electrons start getting pretty close to each other, then the Pauli principle begins to make itself felt. It turns out that whereas a classical gas can shrink indefinitely and become arbitrarily hot, the Pauli principle puts a ceiling on the temperature a given mass of gas can acquire by contracting. So, there is a maximum temperature $T_{max}$ a given mass of gas can have. Obviously, there is also a minimum size.

A hot gas cloud is not necessarily a star. A true star is one where there is burning inside—nuclear burning in fact. So the question now is: Is $T_{max}$ high enough for nuclear burning to take place? Knowing the nuclear ignition temperature, one can then ask: What is the smallest size a star can have? Next, what about the maximum size? Weisskopf asks all these questions and answers them. In between, he also manages to squeeze in some estimates about the largest size possible for a planet as well as the smallest size. Finally, he speculates on how stars die. Let us now go through all this in slightly greater detail.

## 7.3 The virial theorem

There is an important theorem in classical mechanics called the *virial theorem* (see Box 7.1). Applied to the present case, it relates the gravitational energy of the gas to the work needed to compress it from a state of infinite dilution to say, a spherical volume. In practical terms, we have

$$PV = -(1/3) \text{ gravitational energy} \qquad (7.1)$$

Now gravitational energy of a sphere can be written as (see Box 7.2)

---

**Box 7.1** In classical mechanics, one often studies the motion of particles under the action of forces. Usually one considers just a few particles—frequently just one, sometimes two, and occasionally three. However, there are times when one has to deal with a large number of particles, as in a gas. In such cases, there is a theorem called the *virial theorem* which comes in handy.

Let there be $N$ particles, which we assume move in a line, i.e. the motion is one-dimensional. The position of the $i$th particle is $x_i$ and the force acting on it is $F_i$. Now define

$$C = \langle \sum_i F_i X_i \rangle$$

To obtain $C$, we first multiply each $x_i$ by the corresponding $F_i$. We do this for every $i$ and sum over all of them. Now $x_i$ and $F_i$ do not remain constant in time; they would vary. To allow for fluctuations and variations, one now observes the system for a sufficiently long period, computes the above sum at various instants during this interval, and then calculates the average. The angular brackets $\langle \rangle$ denote such an average. The resulting quantity $C$ is called the *virial of Clausius* (actually, it is $C/2$ which is the virial, but never mind).

What do we do with this virial? This is where the virial theorem comes in. It says the average kinetic energy is equal to minus one-half the virial, i.e.

$$\langle K \rangle = -C/2$$

Suppose now the forces $F_i$ acting on the particles are of the inverse square type, i.e. they vary as $(1/r^2)$ (note: this is the case with the gravitational force). One can then deduce

$$\langle K \rangle = - \langle P.E. \rangle /2$$

where P.E. is the potential energy. In the case of a gas, it turns out that

$$PV = -C/3$$

Therefore,

$$PV = (2/3) \langle K \rangle = - \langle P.E. \rangle /3$$

We use this result in the text.

---

**Box 7.2** Consider a sphere of radius $R$, made up of matter of uniform density $\rho$. Every part of the sphere would attract every other part, and so there would be a certain amount of gravitational potential energy associated with the sphere. This energy $\Omega$ is given by

$$\Omega = - (3/5) G(M^2/R) \qquad \text{(a)}$$

If you know some calculus, this result is quite easy to derive. Consider first a core of radius $r$ and a shell of thickness $dr$ around it. Calculate next the attraction between the two; finally, integrate over $r$ from 0 to $R$. Result (a) would follow.

$$\Omega = -(3/5)(GM^2)/R_s \tag{7.2}$$

Using this result in (7.1), we get

$$P \sim GM^2/(VR_s) \tag{7.3}$$

Everything now depends upon what formula one plugs in for the pressure $P$. The game therefore is: (i) Make a reasonable guess for $P$. (ii) Use this in (7.3). (iii) Then compute $R_s$ the radius, as a function of the mass $M$.

### 7.3.1 Classical gas
We suppose that the classical gas law is obeyed. For one mole of the gas we have $PV = RT$. If there are $N$ particles in the volume $V$, then we have instead

$$PV = Nk_BT \tag{7.4}$$

Therefore,

$$P = (Nk_BT/V) \sim (GM^2/R_sV) \tag{7.5}$$

or, dropping factors of the order of unity,

$$k_BT \sim GM^2/(R_sN) \tag{7.6}$$

Now

$$M = N(m_p + m_e) \sim Nm_p \tag{7.7}$$

Further,

$$(R_s)^3 \sim Nd^3 \tag{7.8}$$

where $d$ is the average spacing between adjacent protons. Therefore,

$$k_BT \sim (Gm_p^2 N^{2/3})/d \tag{7.9}$$

Introducing the gravitational fine structure constant $\alpha_G$ defined by

$$\alpha_G = (Gm_p^2)/\hbar c \tag{7.10}$$

and $N_0$ defined by

$$N_0 = (1/\alpha_G)^{3/2} \tag{7.11}$$

eqn (7.6) can be written as

$$k_BT \sim (N/N_0)^{2/3}(\hbar c/d) \tag{7.12}$$

Shortly we shall see that $N_0$ is coincidentally, approximately equal to the number of protons in the Sun. What eqn (7.12) says is that given a gaseous cloud with $N$ particles and obeying the classical gas laws, the temperature increases as the cloud becomes smaller, i.e. as $d$ decreases. Though pressure tries to distend the gas cloud, the classical gas pressure is not strong enough to fight gravity; gravity wins.

An interesting aside. When the gaseous cloud contracts, its temperature increases. This we have already noted. Let us now look at the total energy. This is the sum of the kinetic and the potential energies. On account of the virial theorem, the total energy $U$ is given by

$$U = (1/2)\, \Omega = -(3/10)\, GM^2/R_s \tag{7.13}$$

So when the cloud shrinks, the total energy decreases. Actually the numerical value of $GM^2/R_s$ increases but the negative sign means that the actual value of the energy decreases. So we have a system which is losing energy but is becoming hotter! This is really not as puzzling as it might appear. Remember the total energy is made up of the potential plus the kinetic energies. When $R_s$ decreases, the kinetic energy increases as a result of which the temperature also increases. However, at the same time, the potential energy decreases; and it decreases faster than the increase of the kinetic energy. So the net energy decreases.

Weisskopf gives a good example to drive home the point. He says:

> A similar effect can be seen in the case of the motion of the Moon around the Earth. Suppose we try to reduce the speed of the Moon by hitting it with a large rocket in the opposite direction to its motion. After the drastic blow, its (i.e. the Moon's) energy would be decreased and it would start to fall towards the Earth. Eventually it would take up an orbit nearer to the Earth, and in that case its orbital velocity would actually be greater than before. So if you try to brake the Moon's speed, it goes faster!

**7.3.2 Electrons want space; Pauli supports them!** There is a major flaw in the reasoning of section 7.3.1, which is that when electrons come close to each other, the classical gas law totally fails since the Pauli principle now becomes important. Remember that the Pauli principle leads to the degeneracy pressure. The magnitude of this pressure depends largely on how close the electrons are to each other, and the pressure exists even at zero temperature (unlike the classical gas pressure).

Weisskopf says that we must modify (7.4) as

$$PV = Nk_BT + (N\hbar^2/m_e d^2) \tag{7.14}$$

As a result eqn (7.12) becomes modified as

$$k_BT + (\hbar^2/m_e d^2) \sim (N/N_0)^{2/3} (\hbar c/d) \tag{7.15}$$

or

$$k_B T \sim -A/d^2 + B/d \tag{7.16}$$

where $A$ and $B$ are suitable constants.

**7.3.3 Small hot star** Let us plot (7.16) as also (7.12) and compare the two (Fig. 7.1). The following facts emerge.

1. For a given value of N, the star has a maximum temperature given by

$$k_B T_{max} \sim (N/N_0)^{4/3} m_e c^2 \tag{7.17}$$

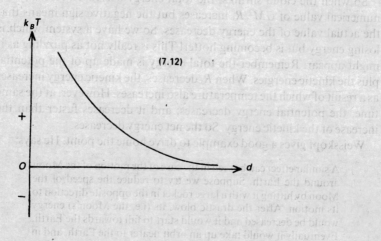

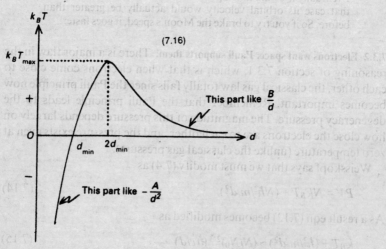

**Fig. 7.1** Schematic plots of eqns (7.12) and (7.16).

2. As $d$ is decreased below the value corresponding to $T_{max}$, the temperature starts dropping. You may wonder how $d$ can be further decreased. Will not the Pauli principle come in the way? Actually, Pauli pressure decreases slightly when temperature is reduced. One can take advantage of that. In fact, the temperature can be brought down to 0 °K. The star is now a cold hunk of matter. The lowest value of $d$ attainable corresponds to this temperature, and is given by

$$d_{min} \sim (\hbar/m_e c)(N/N_0)^{2/3} \qquad (7.18)$$

3. The value of $d$ corresponding to $T_{max}$ is about $2\,d_{min}$.
4. The value of $d$ cannot be reduced any further below $d_{min}$ since that would mean the star has negative temperature which is impossible.

When the temperature decreases from $T_{max}$ to 0, the star radiates away all its heat in a very short time. During this stage the stars are called *white dwarfs*—more about that later.

### 7.3.4 Smallest size possible at $T_{max}$

What is the smallest size a star can have at $T_{max}$? Before we answer this question, we must note that so far we have been considering merely a hot gas and not a burning gas. A star is a star only when nuclear burning takes place inside it (see Box 7.3). It is known from laboratory experiments that nuclear reactions can never take place if the energies of the colliding particles are less than about 0.1 MeV (each). Remembering that the rest mass of the electron corresponds

---

**Box 7.3** You must be familiar with chemical reactions like

$$2H_2 + O_2 = 2H_2O \qquad (a)$$

You must also be aware that chemical reactions are of two types, *exothermic* and *endothermic*. In the former, heat (i.e. energy) is liberated while the latter requires energy input to make it go.

Nuclear reactions are very similar to chemical reactions, except that the participants are nuclei rather than atoms. A typical exothermic nuclear reaction is

$$_1H^3 + {}_1H^2 \rightarrow {}_2He^4 + {}_0n^1 + energy,$$

i.e.

triton + deuteron → alpha particle + neutron + energy

Typically, the energy released in a nuclear reaction is a million times the energy released in a chemical reaction. This reflects the difference between the

> scales of nuclear and chemical binding energies. Whereas chemical binding energies are of the order of eV, nuclear binding energies are of the order of MeV. Incidentally, the nuclear reaction I have given above as an example is the one used in the hydrogen bomb!

to an energy of about 0.5 MeV, Weisskopf supposes that nuclear reactions will not start unless the energy is at least about $fm_ec^2$ where $f$ is $\sim 0.045$. So,

$$k_B T_{max} > fm_e c^2$$

But from (7.17) we have

$$k_B T_{max} \sim (N/N_0)^{4/3} m_e c^2$$

Therefore

$$(N/N_0) > f^{3/4} \sim 0.1 \tag{7.19}$$

Now $N_0$ is of the order of $10^{57}$ which, by coincidence, is also the number of protons in the Sun. So the above relation says that for nuclear reactions to take place, i.e. for the star to be a star, $N$, the number of protons in the star, must be at least one-tenth the number in the Sun; otherwise, no star.

This result should not disturb us for it allows room for objects other than stars to exist in the Universe, our Earth, for example.

**7.3.5 When no star?** When does matter *not* become a star? Weisskopf answers this question in an interesting way. Naturally, he takes a clue from our own Earth. Here there is no burning, no radiation, no nuclear reactions and no compression of matter to the degeneracy limit. The last point is very important. So what kind of matter do we have here? We have atoms such as we studied about in an earlier chapter. And these atoms are well spaced apart and not crushed together.

We have already noted that the size of an atom is about one Bohr radius. So we demand (we pretend as before that matter is made up of only protons and electrons; this is not too drastic an assumption for our present purposes)

$$d_{min} > a_0 = \hbar^2/m_e e^2 \tag{7.20}$$

This tells us that

$$(\hbar/m_e c)(N_0/N)^{2/3} < \hbar^2/m_e e^2 \tag{7.21}$$

*Something about stars* 89

In other words,

$$N/N_0 < (e^2/\hbar c)^{3/2} \sim \alpha^{3/2} \sim 1/1000$$

Thus, for matter as we know it on this earth, we must have

$$N < N_0/1000$$

What the above result says is that if we must have an object like one of the planets in the solar system (remember, planets are not stars.), the mass of the object must be less than 1/1000 the mass of the Sun. In other words, this indicates the largest size a planet can have. Jupiter is in this category.

### 7.3.6 Smallest planet possible

What is the smallest size a planet can have? This is a problem of finding the lower limit. Once again, Weisskopf does this by a clever method. He says: A planet must after all be reasonably spherical. Why? Because it is generally formed by the accretion (i.e. gathering together) of 'cosmic dust'. Moreover, the process is accompanied by spinning. All these forces automatically 'mould' the planets to an approximately spherical shape. For example, who ever heard of a planet that is like a long rod? OK. If a planet is to be spherical, then it means the radius of the planet cannot be smaller than the height of the largest mountain that can exist on the planet. We have already calculated this height for any celestial object containing $N$ atoms. Let us use that result.

Now from eqn (5.20) we have

$$h_c/a_0 = (0.1 \times \eta) f^2 (\alpha/\alpha_G) (1/N)^{1/3} (1/A)^{5/3}$$

where $N$ is the number of nucleons in the planet. As in (5.13), we can write the radius of the planet as

$$R \sim (fa_0)(N/A)^{1/3} \tag{7.22}$$

By Weisskopf's criterion, for the smallest planet,

$$R_{min} \sim h_c$$

Thus using (7.22), we get

$$R_{min} \cdot R_{min} = R_{min}^2$$
$$= (0.1 \times \eta) f^3 a_0^2 (\alpha/\alpha_G)(1/A^2) \tag{7.23}$$

Observe that $N$, the number of nucleons in the planet, does not occur in (7.23). It is to get rid of this $N$ that Weisskopf calculates $R_{min}^2$ rather than

$R_{min}$. Result 7.23 is in terms of fundamental constants alone, barring $A$ which says something about the planet material.

If we substitute numbers, this minimum radius works out to be around 500 km. So no planet can have a size smaller than about 500 km. You have probably heard about asteroids; if not find out. The largest asteroids are about this size.

You may wonder: But I can have a stone which I can hold in the palm of my hand. Is there not something wrong with the above estimate? Is there? Think about it.

## 7.4 Some further thoughts

Weisskopf's discussion on stars was based on the notes of some lectures given by the physicist Salpeter. I shall take this opportunity to say something more about stars, drawing from the work of Chandrasekhar (for the full story, see the companion volume, *Chandrasekhar and His Limit*).

Chandrasekhar's most famous work is on stellar evolution. If a star is a mass of burning gas, then what happens to the star when the fuel is exhausted? In other words, how do stars die? You think this is a weird question? Well, let me tell you that spectacular discoveries, mind boggling in fact, have been made while trying to find answers to this question. What I intend to do now is to briefly narrate the first chapter of that story.

We now know that stars can die in three different ways, leading to three differnt types of stellar 'corpses'. These are the *white dwarf* (see Box 7.4), the *neutron star/pulsar*, and the *black hole*. Of these, it was the white dwarf which was understood first; and Chandra had much if not everything to do with it.

---

**Box 7.4** The star most familiar to us is of course the Sun. It is a flaming mass of gas in which a fierce nuclear fire is raging. The same is true of many other stars. What happens when the fire dies? In the case of a coal fire, we know that what remains at first are the hot cinders which glow and radiate heat. Eventually they cool off and only ashes are left. What is stellar ash like? Everyone is agreed that it would contain electrons, protons and may be a few other nuclei. Also that this matter would be at very high density. Such matter does not exist on Earth but it surely does out there in space where there are many dead stars.

Walter Adams was the first one to discover such a star. There is a bright star in the sky called Sirius. Actually it is a double star, meaning that Sirius has a companion (see figure). For a long time people suspected that Sirius had a companion but it was Adams who confirmed it. While the mass of Sirius is about the same as our Sun, the companion is about 30 times as massive; but its radius is only aroud 20,000 km (for comparison, the radius of the earth is about 12,000 km) which makes this companion a very very dense fellow. Soon after the discovery of this companion, Eddington said, 'Adams has confirmed the suspicion that matter 2000 times denser than platinum is not only a possibility but actually present in the Universe.' What Fowler did thereafter was to try and write down an equation of state for the kind of matter that made up stars like the companion of Sirius. These kind of stars are called *white dwarfs*. They are the remains of stars which have stopped burning.

After the burning is over, there is no pressure generated by the intense heat associated with burning, and so the star starts collapsing. It is this which causes the star to remain hot and shine. But the temperature is no longer all that high. Of course, the collapse does not go on for ever; pressure again stops it, and this time it is the degeneracy pressure. In a nutshell, a white dwarf is a star in which nuclear burning has come to an end. The pressure which sustains the star against gravitational collapse is degeneracy pressure and not the pressure generated by the heat of nuclear reactions.

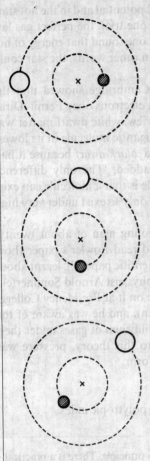

Different views of a double star.

The story goes somewhat as follows: Around 1926, the famous astronomer Eddington did some calculations and argued that despite the very high density, matter corresponding to a white dwarf would obey perfect gas laws. This of course does not make sense, but at that time everybody believed it must be this way. However, this resulted in the following paradox: It is known that stars massive compared to the Sun exhaust their fuel quite rapidly and burn out. There is then nothing left for them but to cool off like a dying fire does, and stars do cool. This means that energy of cold gas is less than the energy of hot gas. In the cold state, the energy of a classical gas is all potential and in the hot state it is largely kinetic. On the other hand, if one took the perfect gas law very seriously and applied it to large stars, one found that energy of hot state is less than the energy of cold state! In other words, the star could not cool which of course is odd.

Shortly afterwards, R. H. Fowler of Cambridge showed that the paradox disappears if one remembers that electrons obey Fermi–Dirac statistics (see Box 7.5). According to this view, white dwarf matter was not like an ionized gas but rather like a 'gigantic molecule in its lowest quantum state'. Fowler called this object a *black dwarf* because it had no energy to give off as radiation. He added, 'The only difference between black dwarf matter and a molecule is that a molecule can exist in a free state while the black dwarf matter only so exist under very high pressure'.

Around 1929–1930, Chandra, then a young man of about twenty, became interested in this business. He had read Fowler's paper (how many college students ever read original scientific papers?), learnt about Fermi–Dirac statistics (the great German physicist Arnold Sommerfeld visited India in 1928, and he gave a lecture on it at Presidency College, Madras where Chandra was a student then), and he was aware of the theory of polytropes, i.e. the theory of equilibrium of gases under their own gravitational attraction. According to this theory, pressure was related to the density $\rho$ by a relation of the form

$$P = K\rho^{(1+1/n)}$$

where $K$ is some constant and $n$ is called the polytropic index.

---

**Box 7.5** Earlier I introduced you to the Pauli principle. There is a practical consequence of this principle when one deals with a large collection of electrons (or protons or neutrons for that matter). Let us suppose that the energy levels of a single electron are as in figure (a). We further suppose that every electron in the collection has the same energy level structure.

Let us now try to populate these levels with electrons. The two rules we must follow are: (i) no level can have more than two electrons, one with spin-up and the other with spin-down, and (ii) the total energy of the entire assembly must be kept as low as possible.

Let us now implement these rules. It is clear we must start from the lowest energy level, put two electrons in it, then go to the next higher level, put two electrons in it, go to the next higher one and so on till all the electrons are accommodated. The occupation pattern would then be as in figure (b) which could be graphically represented as in figure (c).

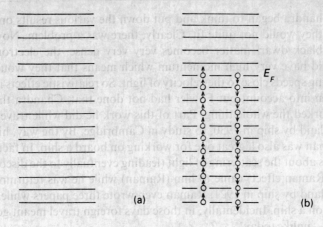

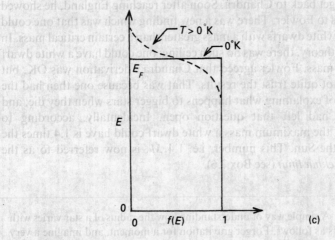

The situation in figure (c) corresponds to 0 °K. At finite temperatures, the occupation probability graph would get modified as shown by the dotted line. If $f(E)$ denotes the probability of a level of energy $E$ being occupied at temperature $T$, then $f(E)$ is given by

$$f(E) = 1/\{[\exp(E - E_F)/k_B T] + 1\}$$

The quantity $E_F$ is called the Fermi energy. The above formula was first derived by Enrico Fermi, and is referred to as the Fermi–Dirac distribution. Notice that even at 0 °K, electrons can have kinetic energies upto a maximum value of $E_F$. By contrast, in a classical gas the particles would have zero kinetic energy at 0 °K. It is this which enables a Fermi gas to exert pressure even at the absolute zero of temperature.

Chandra began to think and put down the various results on paper; but they would not quite fit. Clearly, there was a problem. Now when the black dwarf matter becomes very very dense, the electrons in it would have very high momentum which means that they would start having speeds close to the velocity of light. So relativistic effects must be taken into account but Fowler had not done that. Chandra therefore reworked the whole thing. Part of this work he did while travelling to England by ship in 1930, to study in Cambridge. By the way, his uncle Raman was also a great one for working on board a ship. In fact, many ideas about the scattering of light (leading eventually to the discovery of the Raman effect) came to him (Raman) while he was returning from England by ship in 1921. Raman even wrote three papers while travelling on a ship. Incidentally, in those days foreign travel meant going by ship—unlike today.

Let us get back to Chandra. Soon after reaching England, he showed his results to Fowler. There was a new finding which was that one could not have white dwarfs with a mass greater than a certain critical mass. In Fowler's theory, there was no such ceiling; one could have a white dwarf with any mass. Fowler agreed that Chandra's derivation was OK, but still did not quite trust the results. That was because one then had the problem of explaining what happens to bigger stars when they die; and Chandra had left that question open. Incidentally, according to Chandra, the maximum mass a white dwarf could have is 1.4 times the mass of the Sun. This number, i.e. 1.4 $M_\odot$ is now referred to as the *Chandrasekhar limit* (see Box 7.6).

---

**Box 7.6** A simple way of understanding how the radius of a star varies with mass $M$ is as follows. Forget gravitation for a moment, and imagine a very large gaseous cloud extending upto 'infinity'. Next, suppose the cloud is compressed into a sphere of radius $R_s$ and pressure $P$. Naturally, a certain amount of work has to be done for this purpose. Work means energy, and this is the energy which must be externally supplied if the gas has to be maintained as a sphere of radius $R_s$.

We now 'switch on' gravity, meaning we leave the job of 'squeezing' to gravity, rather than doing it ourselves. In other words, the work required for compression equals the gravitational energy of a cloud of radius $R_s$. Equating these quantities gives the result

$P = (\text{const.}) M^2 / R_s^4$

Different assumptions for $P$ give different results for the Mass–Radius equation.

Obviously, the first thing to try would be the perfect gas law

$$PV = Nk_BT$$

or

$$P = Nk_BT/(4\pi R_s^3/3)$$

This gives

$$R_s \sim (M/T)$$

Classical gas laws are of course not valid in the regime of interest. That is why Fowler decided to use the expression for $P$ as given for a Fermi gas. This resulted in

$$R_s \sim (1/M^{1/3}) \tag{a}$$

According to this, a white dwarf with any mass is possible—only, the larger the mass the smaller the radius.

Chandra observed that if $R$ was small, the electron momentum would be very high which means that electrons would be travelling close to the speed of light. So pressure must be calculated for a *relativistic* Fermi gas and not a nonrelativistic gas as Fowler had done. Chandra took care of this and came up with the approximate result

$$R_s \sim M^{2/3}\{1 - (M/M_\odot)^{2/3}\}^{1/2} \tag{b}$$

The $M$ versus $R_s$ curves for (a) and the exact result of Chandra are shown in the figure. We see that Fowler's result is OK for stars with a small mass, but for larger masses, Chandra's result predicts something dramatically different. There is no stellar corpse possible with mass exceeding 1.4 $M_\odot$ for such an object would have negative radius which of course is not allowed.

Eddington couldn't believe such a prediction, and so tried to pooh-pooh Chandra's amazing result. But now we know he was wrong and that Chandra was right. Indeed, white dwarfs with mass greater than 1.4 solar masses are not possible.

Whatever happens to massive stars? How do they die? What are their corpses like? Answers to these became known only recently, and we know now that heavier stars end their lives in a very spectacular fashion. But that is yet another story. Going back to Chandrasekhar and his discovery. It transcends astrophysics. It revealed that like the familiar law $PV = RT$ of classical physics, there was a new equation of state for dense degenerate matter in which particles are relativistic. Chandrasekhar further observed that matter obeying such an equation of state exists in the interior of certain stars, and that his equation had deep implications for how stars die. When he first wrote

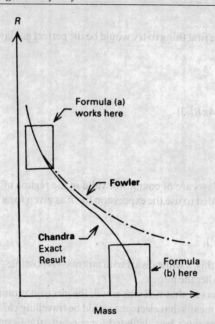

this down he was ignored. But he was correct and fame eventually came to him, though perhaps somewhat late. By the way, when white dwarfs cool off completely, they would become black dwarfs, the objects Fowler had speculated on. It is believed the cooling time required is quite large, and that no white dwarf has had enough time to become a black dwarf yet.

In January 1935, there was a meeting of the Royal Astronomical Society in London. Chandra presented his results, and as soon as he sat down, Eddington got up and declared that Chandra's results were all wrong. Later, Chandra appealed to Milne another leading astronomer but he too would not accept Chandra's theory. In fact he wrote:

> Your marshalling of authorities such as Bohr, Pauli, Fowler, Wilson, etc., very impressive as it is, leave me cold. If the consequences of quantum mechanics contradict the obvious much more immediate considerations, then something must be very wrong with the principles underlying the equation of state derivation.

This is incredible. The year is 1935, and quantum mechanics had been around for many years by then. Even chemists had begun to accept it and use it. Yet some leading astronomers thought it was shady. Chandra was disheartened no doubt but stuck to his guns. As the years rolled by, people realised that Chandra was right and Eddington wrong. Finally, almost half a century later, he was rewarded with the Nobel Prize.

Let us now briefly turn to the question of what happens to stars when their mass exceeds the Chandrasekhar limit. We now know that, depending on how much their mass is in excess of this limit, they end their lives either as neutron stars or as black holes. Black holes are perhaps the most exotic objects in this Universe. They require a full book for description, which I shall probably write later.

Time to wind up this chapter, and let me do it by going back to Weisskopf and quoting him.

> On Earth, atomic physics is important, the temperatures being in the electron-volt regions. In the centres of stars, nuclear physics is important because the temperatures are in the million electron-volts region. We now ask: 'Where in the Universe is high-energy physics at home in the sense that it is the main process?' The answer lies in the fact that the only source available for producing bulk energies is gravity. . . High-energy phenomena will occur when particles have kinetic energies comparable to their rest masses. . . The exciting possibility arises that high-energy phenomena in bulk are really closely connected with gravitational phenomena, thus providing a tool for our investigation of the extreme conditions of stellar evolution. There is the link of the infinitely small to the infinitely big.

And in this way, Weisskopf finally reveals to his student audience, the larger significance of the work going on at CERN.

# 8 Cosmic Numbers

I am now going to discuss something which Weisskopf did not touch upon in his lectures. This material is included for two reasons: Firstly because this too deals with fundamental constants and secondly, this also involves speculation/estimation.

We have already discussed the fundamental constants in Chapter 2. We shall now consider some special groupings of them which are all dimensionless. The constants of interest are:

$$
\begin{array}{ll}
\text{(i) } m_p/m_e & \text{(ii) } e^2/\hbar c \\
\text{(iii) } e^2/Gm_p^2 & \text{(iv) } e^2/Gm_p m_e \\
\text{(v) } \hbar c/Gm_p m_e & \text{(vi) } e^2/Gm_e
\end{array}
\qquad (8.1)
$$

If we substitute numbers and plot them on a logarithmic scale (Fig. 8.1), we find that these dimensionless constants form essentially two groups which are *widely* separated from each other. Notice that the values of the members of one group are very large (let us call this the set of numbers of order $N_1$), while those of the other are quite small (of the order of unity). This clustering is quite remarkable. Does it have any significance? Let us see.

Some people have argued as follows: 'Listen, you have formed some combination of fundamental constants and plugged in values for them, all of which have been measured in the last hundred years or so. You are now claiming a remarkable clustering. How do you know that the fundamental constants had the same value in the early Universe as those you are using now? The clustering which you are now observing could well be accidental, and due to some peculiar time variation of the fundamental constants.'

The argument is reasonable and must be examined. People have in fact done this and come to the conclusion that the fundamental constants have *not* changed with time. How can they say this? Well, there are many clever methods. For example, let us suppose that the fine structure constant has changed with time, i.e. that long ago it had a value different from the one we measure in the laboratory today. Now let us look at the light coming from a very distant stellar object like a

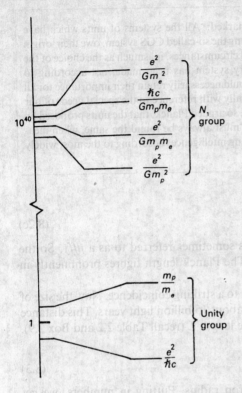

Fig. 8.1 Dimensionless constants of eqn (8.1) plotted (schematically) on a logarithmic scale. Observe the clustering.

quasar. Because the object emitting the light is very far away (and I mean really *very far* away), we would be looking at atoms which existed a very long time ago (remember, light from very distant objects take millions of years to reach us). We can now compare the spectra of those ancient atoms with what we measure today. If the fine structure constant was different, then these spectra would also be different; but they are not. So we conclude that the fine structure constant has indeed remained constant with time. This is typically the way people rule out time variations in the fundamental constants.

Let me interrupt here to introduce you to what may be called the *Planck numbers*. In 1913, Planck showed that the fundamental constants $G$, $c$, and $\hbar$ can be combined to create natural units of length, mass and time (see Box 8.1). The Planck length is

$$a^* = (\hbar G/c^3)^{1/2} \sim 2 \times 10^{-33} \text{cm} \tag{8.2a}$$

The Planck mass is

$$m^* = (\hbar c/G)^{1/2} \sim 2 \times 10^{-5} \text{gm} \tag{8.2b}$$

**Box 8.1** Max Planck has remarked: 'All the systems of units which have hitherto been employed, including the so-called CGS system, owe their origin to the coincidence of accidental circumstances, inasmuch as the choice of the units lying at the base of every system has been made, not according to general points of view which would necessarily retain their importance for all places and all times, but essentially with reference to the special needs of our civilization.' On the other hand, so argued Planck, that the units proposed by him namely, the Planck units, 'must always be found the same, when measured by the most widely differing intelligences according to the most widely differing methods.'

Similarly, the Planck time is

$$t^* = a^*/c \sim 6 \times 10^{-43} \text{sec} \tag{8.2c}$$

The time duration $10^{-43}$ sec is sometimes referred to as a *jiffy*. So the Planck time is about a jiffy. The Planck length figures prominently in theories of the infant Universe.

Time now to draw attention to a striking coincidence. Now the size of the Universe is believed to be about 15 billion light years. This distance is roughly equal to the Hubble length $L$ (recall Table 2.2 and Box 2.9). Let us calculate

$$N_2 = (L/a) \tag{8.3}$$

where $a$ is the classical electron radius. Putting in numbers, we get $5 \times 10^{40}$. Observe

$$N_1 \sim N_2 \tag{8.4}$$

where $N_1$ is the value of the set of numbers shown on top in Fig. 8.1. This coincidence is really remarkable because $N_1$ is derived from the subatomic world, while $N_2$ is derived from the cosmos itself! Is the coincidence accidental or is it meaningful (see Box 8.2)? Before I discuss that, I would like to call your attention to Table 8.1 which says that there is a sequence of cosmic numbers going like 1, $N^{1/2}$, $N$, $N^{3/2}$, $N^2$, $N^{5/2}$,...

**Box 8.2** Dirac made the following observation: 'It is proposed that all the very large dimensionless numbers which can be constructed from the important natural constants of cosmology and atomic theory are connected by simple mathematical relations involving coefficients of the order unity.'

The well-known astrophysicist Martin Rees remarked: 'There are several amusing relationships between the different scales. For example, the size of a planet is the geometric mean of the size of the Universe and the size of an atom; the mass of a man is the geometric mean of the mass of a planet and the mass of a proton. Such relationships, as well as the basic dependences on $\alpha$ and $\alpha_G$ might be regarded as coincidences if one did not appreciate that they can be deduced from known physical theory.'

**Table 8.1** Cosmic Numbers

| Order | Typical ratios |
|---|---|
| 1 | $h^2 H/Gm_p m_e^2 \sim N_1/N_2$ |
| $N^{1/2}$ | $(a/a^*), (L/r), (m^*/m_p), (M/M_\odot)$ |
| $N$ | $(a/a_G), (L/a), (M_\odot/m^*)$ |
| $N^{3/2}$ | $(r/a_G), (L/a^*), (M_\odot/m_p), (M/m^*)$ |
| $N^2$ | $(L/a_G), (M/m_p)$ |

Note:
1. The ratios are of the order indicated, i.e. unity, $N^{1/2}$, $N$, etc.
2. $r = GM_\odot/c^2$ = gravitational size of the star,
   $M$ = mass of the observable Universe,
   $a_G = Gm_p/c^2$ = gravitational size of the proton.

   Other quantities have already been identified.

Let us get back to the mysterious equality of $N_1$ and $N_2$. Several scientists have taken this equality seriously and commented upon it. What do they have to say?

**Dirac:** $N_2$ must vary with time since the size of the Universe increases with time; this of course if you believe that ours is an expanding Universe, which most people do. OK, so $N_2$ varies. Dirac now says that $N_1$ also varies with time, and in this way remains permanently equal to $N_2$.

What about the clustering in Fig. 8.1? If we believe that such a clustering always holds, can we still have a variation of fundamental constants? Dirac says yes. All the constants do not have to vary; it is enough if $G$ alone slowly decreases in time as the Universe expands. If $G$ alone varies, it does not affect the clustering in Fig. 8.1.

Edward Teller (often called the father of the hydrogen bomb) disagrees. He says in effect: 'If Dirac is correct, then $G$ must have had a higher value in the past. Then 3 billion years ago the Earth must have been closer to the Sun than it is now. One can calculate how much closer. One can also estimate what the temperature on the surface of the Earth must then have been. In fact, it would have been so high that nothing could have lived on Earth then. But fossil records show that algae existed 3 billion years ago. So, three billion years ago, the Earth could not have been closer to the Sun than it is now. This means that $G$ could not have decreased with time. If we still believe that $N_1 = N_2$, then we have to look for some other explanation.'

Reference must here be made to the efforts of Gamow to save Dirac's idea. Gamow was one of the colourful personalities of physics, famous for his bold suggestions, simple explanations, and leg-pulling. He also excelled in popular science writing. A famous book of his is called *Mr. Tompkins in Wonderland*. It is about a simple and not very clever bank clerk who falls in love with the daughter of a Professor of Physics. To impress the Professor and win his favour, Tompkins starts attending a course of popular lectures in Physics given by this Professor. But poor Tompkins is not able to follow, soon falls asleep and starts dreaming. Things happen in his dreams, and that is how Gamow skilfully explained atomic structure, etc. Great book.

To get back to what I was saying earlier, if $G$ does not vary, could it be the electronic charge $e$ that varies with time? Gamow thought so (see Box 8.3) but unfortunately he was wrong. I remember an old and somewhat weak Gamow lecturing at the annual meeting of the American Association of Physics Teachers in 1968 in Chicago. Gamow described his theory, and following him came Dyson who literally tore Gamow apart. Gamow was wrong of course but many in the audience felt that Dyson need not have been that harsh. Anyway, Dyson's argument was quite clever. He said: 'The nucleus $_{75}Re^{187}$ decays by radioactivity to the nucleus $_{76}Os^{187}$ (see Box 8.4). If the electronic charge had varied with time even just a tiny bit, then the decay would have been so rapid that there would have been no $_{75}Re^{187}$ left on Earth. But $_{75}Re^{187}$ *is* found on Earth; so $e$ could not have varied with time.' Dyson is right of course.

**Eddington:** He said in effect, 'Who says that $N_2$ increases with time? It appears to because it has been defined wrongly! Define it as

$$N_2 = (c/a\, \Lambda^{1/2}) \tag{8.5}$$

where $\Lambda$ is the so-called *cosmological constant*. If numbers are plugged in, $N_2$ has a value nearly equal to $N_1$.'

This cosmological constant is a story in itself which we will not get into here. Suffices to say here that Einstein introduced this constant when he first formulated the general theory of relativity. Later he called it his biggest mistake. Anyway, for the moment, we shall, like Eddington, accept this constant. What then? Well, then the new definiton of $N_2$ does not change with time and the equality with $N_1$ is always preserved.

**Box 8.3** Reproduced below is the telegram sent by Gamow to his friend and collaborator Ralph Alpher in late 1967. In this telegram, Gamow announces his discovery. The triplet paper refers to a paper by Alpher, Gamow and Herman which was later published in the *Proceedings of the National Academy of Sciences*. The four-inch scar refers to an operation Gamow had just had. Gamow was comparing the length of his scar with one Alpher had acquired during a surgery.

```
WU024 DL PD BOULDER COLO AUG 31 127P MDT
DR RALPH ALPHER R & D CENTER
GE CO SCHDY NY

   TRIPLET PAPER FLIES TO WASHINGTON. LAST NIGHT I MADE EXCITING
DISCOVERY. IF GRAVITATIONAL CONSTANT REMAINS CONSTANT BUT
THE ELEMENTARY CHARGE SQUARED INCREASES WITH TIME,
LUMINOSITY OF THE SUN DEFINED BY CHANGING OPACITY CEDXX DECREASES
VERY SLOWLY WITH TIME AND PRECAMBRIAN OCEANS DO NOT BOIL.
THE ASSUMPTION OF INCREASING ELEMENTARY CHARGE CHANGES OF COURSE
THE RATE OF ALPHA DECAY INFLUENCING URANIUM-LEAD DATING. DUE
TO CHANGE OF RYDBERG CONSTANT SPECTRAL LINES OF DISTANT
GALAXIES SHIFT TOWARD RED, WHICH COMPETES WITH DOPPLER EFFECT. IN
DISTANT FUTURE WHEN FINE STRUCTURE CONSTANT APPROACHES
VALUE ONE, ALL ATOMIC ELECTRONS WILL FALL INTO NUCLEII. SOUNDS
CRAZY BUT SEEMS TO BE CORRECT. DIRAC WILL BE HAPPY. SENDING TODAY
TO PHYSICAL REVIEW LETTERS. MY SCARS ARE ONLY FOUR
INCHES EACH. REGARDS

GEO
```

**The steady-state lobby:** It is generally believed that the Universe is expanding. However, there are some models, according to which, the Universe is in a steady state. The Hubble length for a steady-state Universe is a fixed quantity. So $N_2$ does not vary with time, and the equality between $N_1$ and $N_2$ is always preserved.

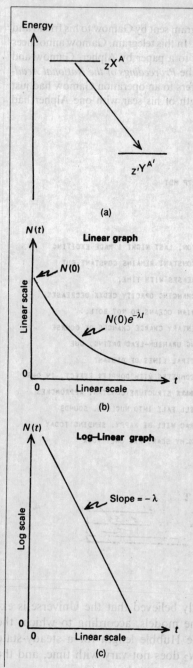

(a)

(b) Linear graph

(c) Log–Linear graph

**Box 8.4** Charge and mass are two of the important properties of a nucleus. Thus it is that one writes $_ZX^A$ while denoting a nucleus. Here Z is the atomic number and denotes the charge, while A is the mass number and indicates the mass.

Radioactivity is the process whereby a nucleus *spontaneously* transforms or decays into another. Symbolically, one can represent such a change by

$$_ZX^A \rightarrow {_{Z'}Y^{A'}} \qquad (a)$$

While A may be (nearly) equal to A', Z is always different from Z' in a radioactive transformation. Nucleus X is called the *parent* and nucleus Y the *daughter*.

A decay like in (a) occurs when nucleus Y has a lower energy than nucleus X—see figure (a). Even though Y may have a lower energy, it does not mean that all nuclei of type X immediately jump down to become nuclei of type Y as in equation (a). The change-over takes a certain time, some nuclei taking the jump first and others coming down later.

There is actually a rule which gives the rate at which they come down called the law of radioactive decay. Suppose you have a collection of nuclei (of type X say) and watch them decaying. Initially you will find many nuclei decaying but as time goes on and the stock of available nuclei of type X starts coming down, the number of nuclei decaying per unit time also will go down. It is somewhat like water coming out of a tap at the bottom of a tank. Initially, when the tank is full, water will flow out at great pressure but as the tank starts becoming empty, the water pressure will also start going down.

Let $N(t)$ be the number of parent nuclei at time $t$, and let $dN$ denote the number of nuclei decaying between times $t$ and $t + dt$. The rule says

$$dN = -\lambda N(t)dt \tag{1}$$

where $\lambda$ is a constant called the *decay constant*. From this one can derive the following law for population decay (see figures (b) and (c)):

$$N(t) = N(0).e^{-\lambda t} \tag{2}$$

where $N(0)$ is the number of parent nuclei at time $t = 0$. The time $t_{1/2}$ required for the population to decrease to one-half of its initial value depends only on $\lambda$, and is given by

$$t_{1/2} = \log_e 2/\lambda = 0.693/\lambda \tag{3}$$

The time duration $t_{1/2}$ is called the *half-life* of the element.

Radioactive decay occurs in the following different ways:

| Type | Form of eqn (a) |
|---|---|
| Electron emission | $_Z X^A \rightarrow {_{Z+1}} Y^A$ |
| Positron emission | $_Z X^A \rightarrow {_{Z-1}} Y^A$ |
| Alpha emission | $_Z X^A \rightarrow {_{Z-2}} Y^{A-4}$ |
| Electron capture | $_Z X^A \rightarrow {_{Z-1}} Y^A$ |

Alpha emission occurs only in heavy nuclei like uranium, for example. Decay usually occurs by one of the other three processes. $_{75}Re^{187}$ decays into $_{76}Os^{187}$ by electron emission (usually called β-decay). By the way, the fact that electrons are emitted during β-decay does not mean that electrons exist inside the nucleus. From where do they come then? Well, as usual, that is a different matter!

**Some tit bits!** Eddington used to come up with interesting numbers. According to a theory of his, the fine structure constant $\alpha$ can be expressed as follows:

$$\frac{1}{\alpha} = [(16^2 - 16)/2] + 16 = 136$$

Later he added unity to the above formula to obtain 137, which is close to the experimental value of 137.036... Of course, people did not take him seriously, and Beck, Bethe and Riseler published a paper in the German journal *Naturwissenschaften* pulling Eddington's leg. The poor editor thought the paper was a serious one, but later, when he realised that he had been fooled, he became enraged. Talking of fooling the editor, there is another interesting story (see Box 8.5).

Eddington also came up with another number which today is called the *Eddington number*. According to him, $N$ the number of protons in the Universe is given by

$$N = (3/2) \times 2^{26} \times 136 \sim 10^{80}$$

It is interesting that Eddington's estimate of the number of protons in the Universe was in a sense anticipated by Archimedes a few thousand years earlier. That too is a story in itself.

---

**Box 8.5**  In November 1975, Professor Hetherington of the Michigan State University in America prepared a paper for publication in the prestigious journal *The Physical Review Letters*. He had done all the research work himself but happened to use the word 'we' to represent the author instead of 'I'. Hetherington showed the manuscript to a friend who said the scientific part was OK but felt nevertheless that the manuscript would bounce since the journal had a strict rule that single-author papers must use 'I' and not 'we'. Now Hetherington did not want to retype the entire paper in order to replace 'we' by 'I'. He solved the problem by adding another name to his. And this is where the story is. Let us hear it from the Professor. He said, 'After an evening's thought I simply asked the secretary to change the title page to include the name of the family cat, a Siamese called Chester, sired one summer by Willard (one of the few unfixed male Siamese cats in Aspen, Colorado). I added the initials F.D. in front to stand for Felis Domesticus and thus created F.D.C.Willard.' In other words, Hetherington's co-author was a cat!

The journal accepted the paper and published it as due to J.H. Hetherington and F.D.C.Willard (see figure). Soon invitations began to be received for Willard to give lectures, address conferences, etc! At this stage the story came out. The cat was out of the bag, so to speak! Everyone was amused, except of course the editor.

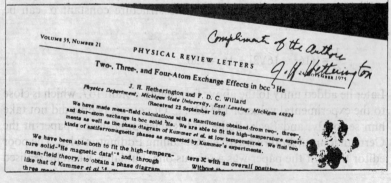

Archimedes did many other things besides the floatation stuff and running out of his bathroom. A remarkable study he made is sometimes referred to as the sand reckoner.

In those days, some kings of Greece believed that our Universe was infinite. Archimedes did not think so and wanted to describe the size of the Universe. So he asked himself the question: How many grains of sand can be packed into the Universe? You may say this was his way of describing the size of the Universe.

Three things are needed before the answer can be attempted. Firstly, one must have a method for writing down large numbers, for surely, the number of sand grains required would come to be a very large number. You may wonder what is so great about writing large numbers. We of course have no problem, thanks to the decimal system invented by our ancients. But the Greeks did not have that system, at least at the time Archimedes lived.

The next thing one must know is the size of the Universe. And lastly, one must know the average size of a sand grain; this of course is not difficult to estimate. Let us now see how Archimedes handled all these problems.

The Greeks knew how to count upto 10,000 which they called a *myriad*. Let us write $m$ for a myriad. Later, the counting was extended upto $m^2$. Archimedes extended it even further. In his system, numbers were expressed as

$$pm^{2[(q-1) + (r-1)m^2]}$$

where $p$, $q$ and $r$ are integers in the range 1 to $m^2$. With this scheme, pretty large numbers could be represented.

Next comes the question of the size of the Universe. Now there was an astronomer called Aristarchus. Based on what he said about the Universe, Archimedes assumed that

$$\frac{\text{size of Universe}}{\text{size of solar system}} = \frac{\text{size of solar system}}{\text{size of Earth}}$$

The size of the Earth was known, and so also the size of the solar system. Using the above relation, the size of the Universe could be calculated. By the way, distances were measured in units of *stadia*. One stadium is about 200 metres.

Now comes the last part, namely the size of a sand grain. Archimedes started with a poppy seed. He assumed that a poppy seed has a diameter equal to 1/10 of the finger-breadth, and further that one poppy seed could contain 1 myriad grains of sand.

In this way, Archimedes found that the Universe (as known to him) could hold $10^{63}$ grains of sand. This is a remarkable number for the following reason. Now the radius of the Universe as we know it today is about $10^{10}$ times larger than that of the Aristarchean Universe; the Universe will therefore hold $10^{93}$ grains of sand. But our Universe has a density $10^{-30}$ times that of the density of sand. So, the Aristarchean Universe filled with sand contains the same mass as the modern observable Universe.

One can estimate that a grain of sand would contain $10^{17}$ nucleons. So, the Aristarchean Universe filled with sand would contain $10^{63}.10^{17} = 10^{80}$ nucleons or protons if you wish. This is precisely the Eddington number for the modern observable Universe.

Going back to the large numbers we started with you may be wondering: So what does it all mean? Is there an equality like $N_1 \sim N_2$ or is it only apparent? I don't think anyone knows the answer for sure. I presented this chapter just so that you might get a feel for some of the speculation going on, involving the fundamental constants. There is also one more point made a long time ago by Gamow. He said that while explaining physical phenomena, one introduces many constants, e.g. surface tension. The job of theoretical physics, Gamow said, was to explain all these empirical constants in terms of the fundamental constants of Nature. I suppose one cannot argue with that.

# 9    *Some Parting Thoughts*

I mentioned in the first chapter that this book is based on Weisskopf's lectures to visiting school students. I am sure that when the students were taken on a tour of the laboratory, the scientists working there must have explained what they were doing, how they were hunting and chasing elementary particles, etc. But when the students met the Director, the latter began talking about atoms, solids, etc., in short, why things are the way they are. Why did Weisskopf do this? Was he just rambling and telling the students a few things he knew? Far from it; there was a deep plan behind his lectures.

Weisskopf was trying to explain to the visiting students *WHY* they were doing at CERN, whatever it is they were doing—in short, why study elementary particles? Notice the way he approaches the subject. He begins his lectures without making any reference to CERN, or the experiments going on there. Instead, he starts talking about atoms, for he knows that students would have been introduced to a bit of atomic physics in high school. And when he talks about atoms, he is not going into the technicalities; rather, he is interested in giving his listeners a *feel* for the sizes of the atoms, and their energies.

Normally, when one reads a subject, one absorbs a lot of detail but often misses the *spirit* of the whole business. Details are no doubt important but that does not mean one should not have a broad feel for the subject as a whole. Let me give an example. These days, there is a fashionable subject called neural networks. This deals with how the neurons in the brain are linked together and function as a computer. Suppose you asked one of these specialists: 'How much information can the human brain store?' I am sure many of these experts would be foxed. The same with sizes, orders of magnitude of physical quantities, etc. Many people simply lack a feel for them. This is what I want to caution you against.

To go back to Weisskopf and his lectures, after dealing with atoms he turns his attention to solids and liquids and shows how their properties may be understood in terms of those of atoms and molecules. Notice he does not do this at a qualitative level—I mean he does not simply say 'heat is the motion of atoms' and stuff like that. Rather, he tells his audience how to make *quantitative* estimates, using simple arguments and fundamental constants.

From solid and liquids, he moves on to stars, gently reminding his audience in the process that while atomic physics is good enough for understanding the Earth, nuclear and elementary particle physics are needed for understanding the stars and our great big Universe. In the twenty or so years that have passed since Weisskopf gave his lectures, not only has our understanding of both the Universe as well as of elementary particles increased tremendously, but also of the *close* connection between the two. In short, we now appreciate better how the very small and the very big are intimately related. Notice how neatly and gradually Weisskopf leads his visitors to the relevance of CERN's programme, which deals mainly with the very small.

There is a lesson to be learnt here—how to present things. I have often seen scientists explaining to visitors. They merrily rattle away, inflicting all kinds of details while the poor visitor is forced to wear a smile and repeat: 'Very interesting'. Of course, all the time, the visitor is probably bored stiff! This happens because the person who is giving the explanation is not able to put himself or herself into the shoes of the listener. People seldom bother to ask: What is the listener's background? Is the person a specialist? If not, how much does the person know about the subject? Will the person be able to relate to what is being explained? Is the person tired? Will the person be able to stand a lengthy explanation? Feynman, Weisskopf and people like them are masters of the art of presentation. This is something we should learn from them.

Talking of learning, let me go back to the point I made in the beginning, namely, the art of making back-of-the-envelope estimates. That was my primary interest in Weisskopf's lectures rather than in his justification of CERN's activities. I have already drawn attention to the fact that one plays this game essentially using the fundamental constants. I now wish to call your attention to another important fact and that is how to retain what is important and throw away what is not. I mean let us say that something we are trying to estimate has the correct value 5.758. There may be three or four factors which determine the magnitude of this particular quantity. The first two may decide that the quantity has a value between 5 and 6 while the other two may influence the decimal places. In making a back-of-the-envelope calculation, one retains only the first two factors; the others are ignored. One must therefore learn what to retain and what to ignore. This trick is not all that easy, which is why I decided to illustrate by simply repeating a great master.

Good physicists are all the time estimating, especially while engaged

Some parting thoughts    111

in discussions. The funny thing is that nobody ever talks to students about the importance of making estimates and how to go about the business. Certainly nobody ever mentioned it to me or taught it to me. One is expected to just pick it up. And if you haven't, people may ignore you. I don't see why everyone should suffer from this handicap, which is why I decided to write this book.

As I said earlier, making estimates is an art. You can pick it up by staying tuned to it, especially while reading the works of masters. Here I should perhaps mention Sir Neville Mott. He is a grand old man, *still* very active. Mott has written many books, and every one of them is full of such estimates. He has no time for long and complicated derivations; he makes his points with simple calculations. So look out for books by Mott. Because he could quickly come to the essence of the matter, Mott has been able to make numerous discoveries, one of which won for him the Nobel Prize.

Talking of estimates, I am reminded of a problem the great Enrico Fermi is supposed to have given his students. Question: Estimate the number of pianos in Chicago. You may think this is crazy; what has it got to do with physics? Of course it has nothing to do with physics but Fermi's idea (—I believe this question was posed to a group of poets who wanted to learn what science was all about) was to expose people to the art of making estimates. The estimate went something like this:

> The number of pianos cannot certainly be greater than the population of the city since at the most every person will have one piano. Say the population is one million; so the number of pianos cannot exceed one million. But actually, it is too much to suppose that every person would have one piano each. A more reasonable assumption is that every family would have one piano each. Let us say there are 5 members in each family on the average. This means that the number of pianos cannot exceed 200,000. Even this is too much. All families certainly cannot afford pianos, not the poor ones anyway. Let us say only the very affluent can afford. Now let us suppose only 20% of the families belong to this bracket. So the number now comes down to 40,000. But now every rich family need not necessarily possess a piano; maybe only 30% do. The number of pianos then comes down to 12,000.

You may not agree with some of the percentages I have assumed along the line but that is not the point. This example is just to illustrate how one makes arguments, guesses, puts upper limits, lower limits, etc.

One of the burning issues in high-energy physics is whether the proton is stable or it can undergo radioactive decay. This is a story in itself, and maybe I will write about it someday. But here I wish to call attention to an interesting limit put by a scientist named Maurice Goldhaber. He said: 'We know it in our bones that if the proton decays, it must have an average lifetime greater than $10^{16}$ years.' Now how on earth did he pull out a number like that? Very simple. Goldhaber said: 'Our body contains a lot of hydrogen; in fact, I can calculate the number of hydrogen atoms in a person of average size. Let me now assume that the half-life for the radioactive decay of the proton is $10^{16}$ years. This means that roughly 30,000 protons in the body must decay every second. If there is that much of radioactivity in the body, it would be a terrific health hazard and we would all be dead! But we are not; which means that so many disintegrations of the proton are not taking place every second; and that in turn means that the half-life is greater than $10^{16}$ years.' Neat isn't it?

This sort of thing is done all the time. Physicists have, for example, an estimate of the total number of protons in the Universe. They can argue whether some mass is missing, and if so, how much. When you hear about such things for the first time, you may think that all this is wild guessing. Not so; they all have a sound basis, and all these estimates are widely accepted. Physicists are pretty hard-nosed fellows and they don't accept nonsense.

I hope you have got a broad flavour of the whole business. What you must now do is to get into this estimating game yourself. In the beginning it will not be easy. But like in the case of crossword puzzles, once you get some practice it would not only be easy but also good fun. Try it, and build up your own library of estimates. Good luck!

### Suggestions for further reading

It is difficult to offer suggestions for further reading concerning the topic covered in the present volume. The only books I can think of are:

1. Davies, P.C.W. *The Accidental Universe*,
   Cambridge University Press: Cambridge, 1982.

2. Fermi, E. *Nuclear Physics*, University of Chicago Press: Chicago, 1951.

I became aware of the first book after completing the present volume. Its style, treatment and scope is rather like this book. It deals with what our Universe is, its purpose and whether it is an accident of nature. The second one is somewhat advanced, meant for students at the M.Sc. level. However, the book is full of quick estimates, made rather in the same spirit as here. For a general list, see the companion volume *The Many Phases of Matter*.